Advanced Processes for Simultaneous Arsenic and Manganese Removal

Awwa Research Foundation

About the Awwa Research Foundation

The Awwa Research Foundation (AwwaRF) is a member-supported, international, nonprofit organization that sponsors research to enable water utilities, public health agencies, and other professionals to provide safe and affordable drinking water to consumers.

The Foundation's mission is to advance the science of water to improve the quality of life. To achieve this mission, the Foundation sponsors studies on all aspects of drinking water, including supply and resources, treatment, monitoring and analysis, distribution, management, and health effects. Funding for research is provided primarily by subscription payments from approximately 1,000 utilities, consulting firms, and manufacturers in North America and abroad. Additional funding comes from collaborative partnerships with other national and international organizations, allowing for resources to be leveraged, expertise to be shared, and broad-based knowledge to be developed and disseminated. Government funding serves as a third source of research dollars.

From its headquarters in Denver, Colorado, the Foundation's staff directs and supports the efforts of more than 800 volunteers who serve on the board of trustees and various committees. These volunteers represent many facets of the water industry, and contribute their expertise to select and monitor research studies that benefit the entire drinking water community.

The results of research are disseminated through a number of channels, including reports, the Web site, conferences, and periodicals.

For subscribers, the Foundation serves as a cooperative program in which water suppliers unite to pool their resources. By applying Foundation research findings, these water suppliers can save substantial costs and stay on the leading edge of drinking water science and technology. Since its inception, AwwaRF has supplied the water community with more than $300 million in applied research.

More information about the Foundation and how to become a subscriber is available on the Web at **www.awwarf.org.**

Advanced Processes for Simultaneous Arsenic and Manganese Removal

Prepared by:
Yu Jung Chang and **Bryan D. Black**
HDR Engineering, Inc.
500 108th Avenue N.E., Suite 1200, Seattle, WA 98004
and
David Chang and **Dianne Gehling***
Southern California Water Company
1920 West Corporate Way, Anaheim, CA 92801
*While employed by/with SCWC.

Jointly sponsored by:
Awwa Research Foundation
6666 West Quincy Avenue, Denver, CO 80235-3098
and
Southern California Water Company
Anaheim, CA 92801

Published by:

Awwa Research Foundation

American Water Works Association

IWA Publishing

DISCLAIMER

This study was jointly funded by the Awwa Research Foundation (AwwaRF) and Southern California Water Company (SCWC). AwwaRF and SCWC assume no responsibility for the research study reported in this publication or for the opinions or statements of fact expressed in the report. The mention of trade names for commercial products does not represent or imply the approval or endorsement of AwwaRF or SCWC. This report is presented solely for informational purposes.

CONTENTS

TABLES

FIGURES

FOREWORD

The Awwa Research Foundation (AwwaRF) is a nonprofit corporation dedicated to the implementation of a research effort to help utilities respond to regulatory requirements and traditional high-priority concerns of the industry. The research agenda is developed through a process of consultation with subscribers and drinking water professionals. Under the umbrella of a Strategic Research Plan, the Research Advisory Council prioritizes the suggested projects based upon current and future needs, applicability, and past work; the recommendations are forwarded to the Board of Trustees for final selection. AwwaRF also sponsors research projects through the unsolicited proposal process; the Collaborative Research, Research Applications, and Tailored Collaboration programs; and various joint research efforts with organizations such as the U.S. Environmental Protection Agency (EPA), the U.S. Bureau of Reclamation, and the Association of California Water Agencies.

This publication is a result of one of these sponsored studies, and it is hoped that its findings will be applied in communities throughout the world. The following report serves not only as a means of communicating the results of the water industry's centralized research program but also as a tool to enlist the further support of the nonmember utilities and individuals.

Projects are managed closely from their inception to the final report by AwwaRF staff members and large cadre of volunteers who willingly contribute their time and expertise. AwwaRF serves a planning and management function and awards contracts to other institutions such as water utilities, universities, and engineering firms. The funding for this research effort comes primarily from the Subscription Program, through which water utilities subscribe to the research program and make an annual payment proportionate to the volume of water they deliver and consultants and manufacturers subscribe based on their annual billings. The program offers a cost-effective and fair method for funding research in the public interest.

A broad spectrum of water supply issues are addressed by the AwwaRF's research agenda: resource, treatment and operations, distribution and storage, water quality and analysis, toxicology, economics, and management. The ultimate purpose of the coordinated effort is to assist water suppliers in providing the highest possible quality of water economically and reliably. The true benefits are realized when the results are implemented at the utility level. AwwaRF's trustees are pleased to offer this publication as a contribution toward that end.

Walter J. Bishop
Chair, Board of Trustees
Awwa Research Foundation

Robert C. Renner, P.E.
Executive Director
Awwa Research Foundation

ACKNOWLEDGMENTS

This report is the product of a collaborative effort. The authors appreciate the support and leadership provided by AwwaRF's project manager Alice Fulmer and the Project Advisory Committee including Dr. Sun Liang of the Metropolitan Water District of Southern California, Bob Fuerman of the Missouri-American Water Company, and Tom Sorg of the US Environmental Protection Agency.

The authors appreciate the assistance from Mike Gutierrez, Hank Aceves, Dennis Ahlen, and George Haller of Southern California Water Company in site construction and operation activities. Jason Wen with City of Downey, CA provided the vision for this project and was instrumental in its development.

The authors also appreciate the support received throughout this project from Mr. Floyd Wicks, President and CEO, and Ms. Denise Kruger, Sr. Vice President of Operations of Southern California Water Company. Without their unwavering commitment to research, this project would not have reached fruition.

ACKNOWLEDGEMENTS

[illegible]

EXECUTIVE SUMMARY

The project described in this report is an effort to modify, demonstrate, and optimize a treatment process to simultaneously remove both arsenic and manganese. The U.S. Environmental Protection Agency (EPA) established a new maximum contaminant level (MCL) for arsenic at 10 μg/L in January 2002. All community and non-transient non-community (NTNC) water systems are required to comply by February 2006. The treatment process selected for development and demonstration includes oxidation, ferric chloride addition, and filtration. The process can be used to remove arsenic or to simultaneously remove arsenic, iron, and manganese.

RESEARCH OBJECTIVES

The overall objectives of this project were to modify, demonstrate, and optimize an existing treatment process to simultaneously remove both arsenic and manganese. A secondary objective was to establish an efficient process design approach for other utilities implementing this process. To accomplish these objectives, the existing oxidation/filtration process at the Southern California Water Company's (SCWC) Centralia site was modified to include a ferric chloride feed system and a backwash waste handling system. The following key questions were addressed in the project:

1. Will the addition of ferric chloride to the existing oxidation/filtration process result in simultaneous removal of arsenic and manganese?
2. What ranges of contaminant removal can be achieved?
3. How will filter operations be affected?
4. What are the optimal operating conditions for the SCWC Centralia site?
5. How should treatment process residuals be handled?
6. Can bench-scale testing effectively predict required ferric chloride doses and contaminant removal?
7. Can pilot-scale testing effectively predict required ferric chloride doses, contaminant removal, and filter run volumes/times?
8. What role does filter loading rate play in contaminant removal performance and filter operations?

APPROACH

Iron and manganese concentrations of groundwater at the SCWC's Centralia testing site are 60-85 μg/L and 72-106 μg/L, respectively. The existing conventional oxidation/filtration treatment can effectively control manganese below 20 μg/L, but has very little impact to arsenic concentration.

The approach of this research is to modify the existing oxidation/filtration process to include a ferric chloride feed system as an add-on process to reduce the current arsenic concentration from 16 μg/L to a level below the new MCL. The modified system was operated and monitored under a variety of conditions. Bench-scale and pilot-scale testing were also completed to evaluate their ability to predict full-scale results. The following tasks were completed during this project.

1. **Preparation and coordination**. Key project personnel met to develop the overall project plan, schedule, and treatment objectives.
2. **Bench- and Pilot-Scale Testing**. Bench scale testing was performed at the HDR Engineering, Inc. (HDR) water laboratory to study the impact of ferric chloride dose on filtered water arsenic concentrations. A six-inch diameter pilot filter was operated onsite for comparison to the full-scale system and to evaluate impacts of filter hydraulic loading rate.
3. **Process Design**. Design included a ferric chloride feed system, a static mixer/vault, filter piping modifications, and installation of backwash waste holding tanks. A testing and monitoring plan was developed based on the bench-scale test results.
4. **Construction and Startup.** A contractor installed chemical feed lines, the static mixer/vault, and piping to the backwash waste holding tanks. SCWC and HDR personnel installed the chemical feed system, data loggers, and an electronically actuated decant valve for the backwash waste holding tanks. Startup included verification of chemical feed pump output using a calibration cylinder.
5. **Full-Scale Testing.** Testing included operation and monitoring of the full-scale system under a range of ferric chloride doses. The conditions during each filter run were generally fixed during the run, with operation until iron breakthrough was observed. An automatic sampler was programmed to collect filtered water samples at specified times during the filter run.
6. **Cost Analysis.** Cost curves were developed to estimate capital costs based on design flow rate, and example costs for operation and maintenance (O&M) were developed and presented.

SUMMARY OF MAJOR FINDINGS

1. Addition of ferric chloride to the existing oxidation/filtration process results in adequate arsenic removal without sacrificing manganese removal performance.
2. The implementation of ferric chloride addition is a fairly simple and straightforward process with minimal additional equipment and instrumentation requirements.
3. For the SCWC Centralia site that typically had 16 µg/L influent arsenic, filtered water arsenic concentrations as low as non-detect (<2 µg/L) could be achieved, depending on the ferric chloride dose.
4. A ferric chloride dose of between 2 and 4 mg/L was sufficient to achieve a target filtered water arsenic concentration of 8 µg/L. A ferric chloride dose of 6 mg/L was sufficient to achieve a target filtered water arsenic concentration of 5 µg/L.
5. The addition of ferric chloride shortens filter runs due to the additional solids added, with higher ferric chloride doses resulting in shorter filter runs. At the Centralia site, filter runs were always terminated due to the breakthrough of iron (as opposed to breakthrough of other contaminants or excess headloss). A ferric chloride dose of 4 mg/L was sufficient to reach the target filtered water arsenic concentration of 8 µg/L and resulted in a filter run of 55 hours with a Uniform Filter Run Volume (UFRV) of 8,200 gallons/ft^2.

6. The backwash waste stream did not contain sufficient arsenic concentration to be classified as hazardous waste. However, once concentrated through several batch settling and decant cycles, the retained settled sludge exceeded the 5 mg/L arsenic criterion for liquid residuals. When the liquid waste was dried, it passed the Toxicity Characteristic Leaching Procedure (TCLP) but did not pass California's Waste Extraction Test (WET). Therefore, the solid waste in California would be classified as a non-Resource Conservation and Recovery Act (RCRA) hazardous waste. The solid waste would not have been classified as hazardous outside of California. Due to the residuals characteristics and the lack of space on the Centralia site, the waste can be most readily handled through equalization, settling, decant of the supernatant to the stormwater channel, and discharge of the settled sludge to the sanitary sewer. Costs of the sewer connection were not specifically evaluated but typically include both a connection and use charge with a potential surcharge for strength.
7. Bench-scale testing was found to effectively predict the full-scale testing results for removal of arsenic by ferric chloride and filtration. Bench-scale testing is not expected to effectively simulate manganese removal because its rate of oxidation is impacted by manganese coatings on the filtration media. Bench-scale results do not address filter run lengths or uniform filter run volumes.
8. Pilot-testing effectively simulated the full-scale system. Beyond determining the impact of ferric chloride addition on filtered water arsenic concentrations, pilot testing can effectively predict manganese removal (provided the filter media is seasoned or greensand is used), filter run times, and uniform filter run volumes.
9. Filter loading rates were examined during pilot testing. No correlation was found between filter loading rate and filtered water arsenic concentration for loading rates between 2.2 and 8 gpm/ft^2. The UFRV was slightly lower at higher loading rates. Increasing the loading rate from 2.2 to 8 gpm/ft^2 decreased the UFRV by 10 percent. Because essentially the same volume is produced in the filter run, the filter run length is inversely proportional to the loading rate.
10. Use of backwash water with greater arsenic concentrations than the filtered water was found to artificially increase the arsenic concentrations during the early portions of the filter run. Based on the observation of gradually declining filtered water arsenic concentration at the beginning of the filter run, the filtration media appeared to have a limited capacity for adsorption/desorption of arsenic. Therefore, after exposed to water with higher concentrations of arsenic, the filter media tends to supply dissolved arsenic to the process stream through desorption over the initial eight hours of the run. This should not be a problem for filters that are operated with a uniform ferric chloride dose and are backwashed with filtered water. When backwash water with equivalent or lower concentrations to the filter effluent was used, filtered water arsenic concentrations remained uniform during the filter run.

FUTURE RESEARCH

1. The iron-arsenic solids produced during water treatment and residuals handling behaved very differently when exposed to TCLP conditions as opposed to WET test conditions. Investigations could be completed to determine the cause of these differences and methods of sludge conditioning to pass the WET test. Further, because selection of the

WET test as a regulatory basis will significantly impact arsenic treatment costs in California, the various methods of leach testing should be reviewed for appropriate use as a regulatory threshold.

2. The filter media profile used in this work was fixed at 18 inches of anthracite coal over 12 inches of silica sand. Filter runs were always terminated based on iron breakthrough rather than headloss. A more optimal media design may reach iron breakthrough and terminal headloss simultaneously. Additional work could be performed at the pilot scale to optimize the media profile for longer filter run lengths.

CHAPTER 1
INTRODUCTION

BACKGROUND

The project described in this report is an effort to modify, demonstrate, and optimize an existing treatment process to simultaneously remove both arsenic and manganese. The U.S. Environmental Protection Agency (EPA) established a new maximum contaminant level (MCL) for arsenic at 10 μg/L in January 2002. All community and non-transient non-community (NTNC) water systems are required to comply by February 2006. The treatment process selected for modification and demonstration in this project includes oxidation, ferric chloride addition, and filtration. The process can be used to remove arsenic or to simultaneously remove arsenic, iron, and manganese.

Of the estimated 12 percent of the community water systems in the United States that exceed the MCL, about 80 percent are supplied by groundwater (Frey, Owen et al. 1998). In California, groundwater systems are expected to account for 98 percent of all affected sources and require 95 percent of all the capital spent on arsenic removal facilities (Fruth, Drago et al. 1996). With groundwater wells that exceed the arsenic MCL, the Southern California Water Company (SCWC) represents the many utilities in California and throughout the United States that must take action to achieve compliance with the arsenic rule.

Based on the most recent sampling results from Los Angeles Central and West Basins, there are approximately 50 purveyors exceeding 5 μg/L. Like the SCWC's Centralia groundwater production site, many wells containing arsenic in the Central/West basins also contain elevated concentrations of manganese. Currently secondary MCLs (including 0.3 mg/L for iron and 0.05 mg/L for manganese) are enforceable per California Drinking Water Regulations. The California Department of Health Services (DHS) self-imposed an action level of 0.5 mg/L for manganese, which by definition, automatically put manganese under consideration for regulation as a MCL. With the new arsenic MCL, these affected wells will require treatment processes that can effectively remove both arsenic and manganese.

Like SCWC's Centralia site, many wells in southern California have been equipped with oxidation and pressure filtration for manganese removal. However, this process is not very effective for arsenic removal when iron concentrations in the groundwater are low. There are several known treatment technologies for arsenic removal, including precipitation processes (e.g., coagulation/filtration) and sorption processes (e.g., ion exchange and proprietary adsorbents). Installing a single process to simultaneously remove both arsenic and manganese is desirable to simplify operations, reduce costs, and minimize plant footprint. This report focuses on oxidation, coagulation, and filtration for simultaneous removal of arsenic and manganese.

ARSENIC REMOVAL TECHNOLOGIES

There are numerous technologies that exist for arsenic removal. For systems that do not currently treat water beyond disinfection, the most practical ways to remove arsenic include sorption processes (ion exchange or adsorption) and precipitation processes (coagulation followed by either membrane or granular media filtration). Membrane processes (nanofiltration or reverse osmosis) may be appropriate if constituents other than arsenic also need to be removed (e.g., hardness, total dissolved solids, or nitrate). Considerations for the selection of an appropriate arsenic treatment technology include capital cost, operation and maintenance cost, footprint, waste disposal, and chemical storage issues. Methods to evaluate other arsenic treatment technologies can be found in other Awwa Research Foundation (AwwaRF) research projects (e.g., Awwa Research Foundation 2661, "*Demonstration of Emerging Technologies for Arsenic Removal*"). Arsenic treatment technologies present a challenge for urban groundwater wells due to space, chemical handling, residuals handling, or high cost requirements. Because many well sites in southern California are adjacent to residential areas, small footprint and minimum chemical usage are often required. For the systems that already have pressure filters installed for manganese removal, such as some of SCWC's well sites, using coagulation assisted filtration for arsenic removal would seem to be a rational choice.

PRECIPITATION PROCESSES

Treatment processes that rely on the formation of a metal hydroxide precipitate have the potential to remove both dissolved and particulate arsenic. The steps to remove arsenic by precipitation processes are: 1) ensure that arsenic is in the form of arsenate (add oxidant if necessary); 2) initiate precipitation by adding a coagulant or oxidizing dissolved iron naturally present in the source water; 3) promote the incorporation of arsenic into solids formed during treatment; and 4) ensure effective separation (filtration) of solids from the liquid stream.

Bench- and pilot-scale testing of coagulation/filtration systems has demonstrated very effective arsenic removal. With low to moderate doses of coagulant (particularly iron based), residual arsenic concentrations less than 2 µg/L have been achieved in a number of studies (Amy et al. 2000; Clifford et al. 1998; Chang et al. 1994; Hering, et al., 1996; Hering and Elimelech, 1996). Examples of reported coagulant doses for removing arsenic are shown in Table 1.1. Generally, ferric chloride doses of 3 to 10 mg/L (as $FeCl_3$) are highly effective in removing arsenic.

ARSENIC AND MANGANESE CHEMISTRY

Dissolved arsenic in water occurs in two main chemical forms: arsenite [As(III)] and arsenate [As(V)]. The chemical form of dissolved arsenic, either As(III) or As(V), significantly impacts its removal during water treatment. This is primarily because As(V) is a negatively charged molecule that readily attaches to metal hydroxide surfaces. As(III) is neutral in the pH range of most natural water and does not attach to particles as readily.

Table 1.1
Reported coagulant doses for As(V) removal

Initial As(V) concentration (ppb)	Coagulant	Coagulant dose (mg/L)	Approximate removal efficiency (%)	Reference
0 to 50	Ferric chloride	4.9	95 to 100	Hering, et al., 1996
20	Ferric chloride	4.9	90 to 95	Hering and Elimelech, 1996

If the concentration of As(III) in the raw water is greater than the new MCL, pre-oxidation must be included in the treatment system to convert As(III) to As(V) for subsequent removal. Fortunately, oxidation of As(III) to As(V) occurs extremely rapidly with standard oxidants such as chlorine and potassium permanganate (Ghurye and Clifford 2000). Unless the source water contains exclusively As(V), systems designed to remove arsenic should oxidize the raw water to the treatment process to ensure that all As(III) is converted to As(V).

The geochemistry of arsenic and manganese has been well summarized by others (Edwards, 1994; Welch, 2000). These authors noted that the concentrations of iron and manganese strongly affect the concentration of arsenic found in groundwater. Key minerals that control arsenic solubility include arsenopyrite (FeAsS) and $Mn_3(AsO_4)$. As noted, arsenic can be immobilized by adsorption to iron and manganese hydroxides and mobilized when these solids are dissolved under reducing conditions. Arsenic can also be released from manganese oxide surfaces due to competition for sorptive sites by orthophosphate and natural organic matter (Ferguson and Gavis 1972; Masscheleyn, Delaune et al. 1991; Edwards 1994; Edwards, Patel et al. 1998).

Manganese has also been investigated for use as an adsorbent and solid-phase oxidant in arsenic removal processes (Magyar 1992a; Magyar 1992b; Lauf and Waer 1993; Dreihaus 1995; Lauf and Plsarczyk 1995; Chiu 2000). These studies have examined the kinetics of adsorption and oxidation under of variety of conditions and with a number of forms of manganese oxides including manganite (γ-MnOOH), birnessite (δ-MnO_2), and pyroluscite (β-MnO_2). As noted by Dreihaus (1995), the influence of other commonly occurring reducing species, such as ferrous ions and natural organic matter, may lead to higher concentrations of dissolved arsenic as they effectively reduce As(V) to As(III), which is more readily mobilized. Manganese dioxide filters can oxidize As(III), but removal of As(V) requires a second process such as adsorption coprecipitation/filtration. This project focuses on improving arsenic removal in an oxidation/ filtration process designed to control manganese by adding ferric chloride coagulant.

TESTING SITE DESCRIPTION

The field study was performed at the SCWC's Artesia System Centralia Facility. This facility includes two groundwater wells, Well C3 and Well C4, which have arsenic in excess of the MCL (typically 16 μg/L) and manganese concentrations in excess of the secondary MCL (50 μg/L).

Each well pumps between 500 and 600 gpm and has a flow meter on the pump discharge. Chlorine (sodium hypochlorite) is added to the pump discharge line at a typical dose of 5 mg/L. Wells are operated jointly to pump water through dual media pressure filters and then fill the onsite finished water storage tank. Wells typically operate in automatic mode with an operator onsite about 1 hour per day. The typical water quality and basic operating parameters for these two wells are summarized in Table 1.2. Specific raw water quality testing results from this project are presented in Chapters 3 and 5. Photographs of the wells are presented in Figures 1.1 and 1.2.

Table 1.2
Typical well water quality and operating parameters

Parameter	Well C3	Well C4
Total arsenic (μg/L)	11	16
Dissolved arsenic (μg/L)	11	14
Arsenite (μg/L)	5	6
Manganese (μg/L)	106	72
Iron (μg/L)	85	60
Capacity (gpm)	473	475
pH	7.8	7.9
Conductivity (μS/cm)	454	405
Total dissolved solids (mg/L)	270	260
Hardness (mg/L as $CaCO_3$)	144	129
Alkalinty (mg/L as $CaCO_3$)	190	180
Sodium (mg/L)	42	53
Potassium (mg/L)	2	2
Calcium (mg/L)	46	42
Magnesium (mg/L)	7	6
Chloride (mg/L)	12	14
Sulfate (mg/L)	36	46
Fluoride (mg/L)	0.34	0.32
Phosphate (mg/L)	0.42	0.31
Silica (mg/L)	48	46
Total organic carbon (mg/L)	<0.5	<0.5

Most of the arsenic present in Wells C3 and C4 is soluble. The speciation of well samples show that arsenite, the reduced form of arsenic, comprises about 45 percent of the total arsenic in both wells. Nonetheless, arsenite is readily oxidized to arsenate during manganese oxidation with chlorine and can be removed by filtration as long as it is preceded by an iron-coagulation step. The well water quality in general does not have competing ions (such as phosphate, fluoride, and total organic carbon [TOC]) in the concentration ranges that significantly impair arsenic removal. The concentration of silica (about 46 mg/L) might reduce the arsenic removal efficiency for iron-based technology at high pH (above 8.5), but should not significantly impact its feasibility for this project.

Figure 1.1 SCWC Centralia Well No.3 with hypochlorite building

Figure 1.2 SCWC Centralia Well No.4

Water from the wells is combined in a 8-inch-diameter line and pumped through two dual media Linden Systems pressure filters (Figure 1.3) into the 750,000 gallon Centralia Plant ground-level, welded-steel reservoir that is 36 feet tall (Figure 1.4). Each pressure vessel is divided into two separate filters. Filter media consists of 18 inches of anthracite coal over 12 inches of silica sand (Figure 1.5). Each pressure vessel is 8 feet in diameter, 24 feet long, and supplies a filter surface area of 192 ft^2. The surface-loading rate for the filters is 3 gpm/ft^2 based on the combined maximum pumping rate of the two wells of 1,150 gpm and the total filter surface area of 384 ft^2. An annual inspection of the filter media is conducted to confirm media condition and depth. High service pumps pressurize stored water for distribution (Figure 1.6). A process flow diagram for the system prior to modification is illustrated in Figure 1.7.

Figure 1.3 Pressure filters with inlet/outlet face piping

Figure 1.4 Filtered water storage in ground-level welded steel tank

Figure 1.5 Schematic dual filtration media profile

Figure 1.6 High service pumps

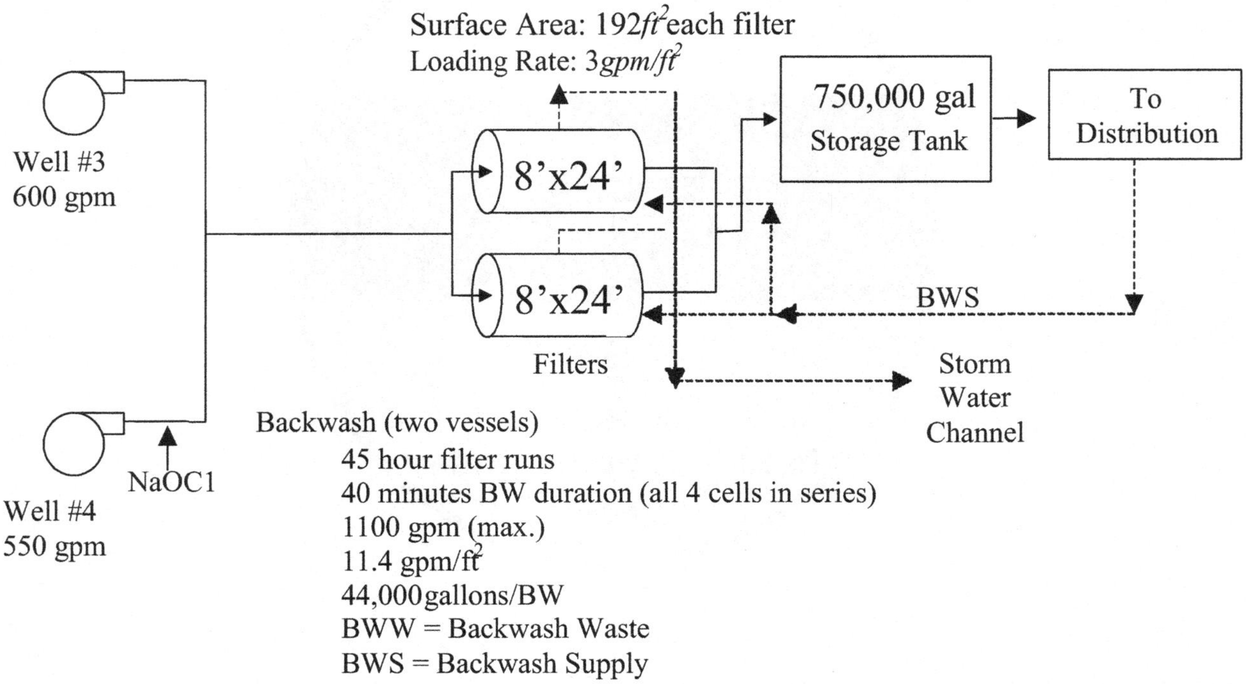

Figure 1.7 Schematic of SCWC Centralia testing site prior to modification

The filters are controlled with a Modicon model 984-120 Programmable Logic Controller (PLC). Prior to modification, filter runs exceeded 100 hours prior to manganese breakthrough. This project reviewed the impact of ferric chloride addition on arsenic removal, other water quality, backwash frequency, filter headloss, and residuals generation and disposal.

The backwash duration is set for 10 minutes per individual filter cell (40 minutes total) at a maximum flow rate of 1,100 gpm (11.4 gpm/ft^2). Backwash and rinse wastewater is discharged to the Los Angeles County Flood Control Channel under an existing National Pollutant Discharge Elimination System (NPDES) permit. The addition of ferric chloride will result in an increase in solids in the backwash waste. Temporary tanks designed for sediment removal from stormwater were leased for the duration of the full-scale system tests to control the quality of water discharged to the stormwater channel.

The existing treatment process typically results in filtered water with less than 0.1 NTU turbidity. pH is typically about 7.8 in the raw water and about 7.5 in filtered water. The existing process effectively removes manganese, with concentrations typically below 0.02 mg/L. But prior to iron coagulant addition, no significant arsenic removal was obtained. Although filter runs under existing conditions could last as long as two weeks, the utility backwashes the filters every 48 hours to manage backwash wastewater quality.

RESEARCH OBJECTIVES

The overall objectives of this project were to modify, demonstrate, and optimize an existing treatment process to simultaneously remove both arsenic and manganese. A secondary objective was to establish an efficient process design approach for other utilities implementing this process. To accomplish these objectives, the existing oxidation/filtration process at the SCWC's Centralia site was modified to include a ferric chloride feed system and a backwash waste handling system.

The following key questions were addressed in the project:

1. Will the addition of ferric chloride to the existing oxidation/filtration process result in simultaneous removal of arsenic and manganese?
2. What ranges of contaminant removal can be achieved?
3. How will filter operations be affected?
4. What are the optimal operating conditions for the SCWC Centralia site?
5. How should treatment process residuals be handled?
6. Can bench-scale testing effectively predict required ferric chloride doses and contaminant removal?
7. Can pilot-scale testing effectively predict required ferric chloride doses, contaminant removal, and filter run volumes/times?
8. What role does filter loading rate play in contaminant removal performance and filter operations?

RESEARCH APPROACH

The following major tasks were performed to investigate arsenic and manganese removal.

- Analysis of existing conditions
- Bench-scale testing to determine coagulant dosage needed to achieve target arsenic removal goal
- Pilot-scale testing to evaluate impact of coagulant to filter run volume or filter run time
- Design and construction of full-scale improvements
- Full-scale testing
- Data management and reports

WATER QUALITY TESTING

Water quality testing was completed by Clinical Laboratory of San Bernardino, California. The lab is certified for drinking water analysis by California Department of Health Services under ELAP Number 1088. Each batch of water quality samples was transmitted to the laboratory under chain-of-custody procedures. Laboratory analytical methods and detection limits are presented in Table 1.3.

The Toxicity Characteristic Leaching Procedure (TCLP) was performed using EPA method 1311 and the California Wet Extraction Test (WET) was performed in compliance with Title 22 regulations. Metals in solids and extracts were analyzed by EPA methods 6010/7000.

The laboratory operates under a QA/QC plan for both Inductively Coupled Plasma Spectroscopy (EPA 200.7) and Graphite Furnace Atomic Absorption (SM 3113B). The quality control plan is available upon request. QA/QC procedures are summarized in Table 1.4.

Table 1.3
Laboratory methods and detection limits

Parameter	Preservative	Method	Detection Limit
Arsenic	HNO3 to pH <2	SM 3113-B	2.0 μg/L
Iron	HNO3 to pH <2	EPA 200.7	100 μg/L
Manganese	HNO3 to pH <2	EPA 200.7	20 μg/L
TSS	Cool 4 C	EPA 160.2	1 mg/L
Silica	HNO3 to pH <2	EPA 200.7	0.50 mg/L
Phosphate	Cool 4 C	EPA 365.2	0.02 mg/L
TOC	HCl to pH <2	SM 5310-B	0.50 mg/L
Fluoride	None	EPA 300.0	0.10 mg/L
Alkalinity	Cool 4 C	SM 2320-B	1.0 mg/L
TDS	Cool 4 C	SM 2540-C	1.0 mg/L

Table 1.4
Laboratory QA/QC procedure summary

Procedure	Frequency
Method blank, liquid control sample, matrix spike, and matrix spike duplicate	Each batch of 20 samples
Linearity check	Each batch of 20 samples
Continuing calibration verifications	Each batch of 20 samples
EPA audit check sample	At least once per year
Method detection limit determination	At least once per year

CHAPTER 2
BENCH-SCALE TESTING

Bench-scale testing was conducted to establish the feasibility of using ferric chloride addition to improve arsenic removal, and to generate an arsenic removal dose-response curve that can be used as the basis for the subsequent pilot and full-scale testing. Testing was conducted at HDR's laboratory in Bellevue, Washington, on raw water collected from the SCWC's Centralia facility. Standard jar testing equipment and protocols were used for this bench study. Arsenic samples were analyzed by a certified laboratory (Clinical Lab, CA). The primary testing conditions are summarized below.

- Run A – Ferric chloride dose varied, pH not controlled
- Run B – Ferric chloride dose varied, pH controlled to ambient (7.89)
- Run C – Ferric chloride dose fixed at 4 mg $FeCl_3$ / L, pH varied
-

General testing procedures for each run included oxidation with sodium hypochlorite, coagulation with ferric chloride, and filtration with 0.45 μm filters. A six-place stirrer was used for chemical mixing. Detailed sampling and testing procedures are included in Appendix A (Bench Test Plan).

RESULTS AND DISCUSSION

Figure 2.1 and Figure 2.2 illustrate that oxidation, ferric chloride addition, and filtration can effectively control finished water arsenic concentrations. The results shown in Figure 2.1 represent the dose response curve with uncontrolled pH conditions (Run A). With the Federal MCL set at 10 ppb, the arsenic treatment goal set by each utility should be less than that in order to provide some provision for operation and water quality fluctuations. In addition to that, the State of California is considering setting a statewide MCL for arsenic that could be lower than 10 ppb. For these reasons, the arsenic treatment goal for this project was set at 5 μg/L. To achieve this goal, Figure 2.1 suggests that 4 to 6 mg $FeCl_3$/L would be required. The bench-scale test results showed that the finished water pH could drop from 7.89 to 7.6 using a ferric chloride dose of 4 to 6 mg/L. This pH drop is not expected to significantly increase corrosion or metal release rates in the distribution system.

Because the addition of ferric chloride reduces water pH, water receiving greater doses of ferric chloride will have lower final pH. However, even at the highest ferric dosage tested in this study (15 mg/L), the pH reduction was only 0.5. This small pH decrease is due to the relatively high alkalinity of the raw water (160 to 180 mg/L).

It should be noted that the filtration of each sample using 0.45 μm membrane filters took more than 5 minutes to complete and the total reaction time for the last sample may be as long as 30 minutes. Although this extended reaction time is much longer than what will be experienced in full-scale operation, the adsorption and pH adjusting process should be completed in the order of several seconds and therefore the extended reaction time should not affect the relevance of the data. The arsenic removal efficiencies observed in the full-scale testing (Chapter 5) match this bench test very well, suggesting the extended reaction time will not enhance arsenic removal. The only reaction that might benefit from a longer reaction time is the oxidation of manganese. However, the manganese results shown in Figure 2.3 indicate that the oxidation of manganese could also be fairly rapid. Nonetheless, the pH change and arsenic removal efficiency at various ferric dosages will be closely monitored in the full-scale operation.

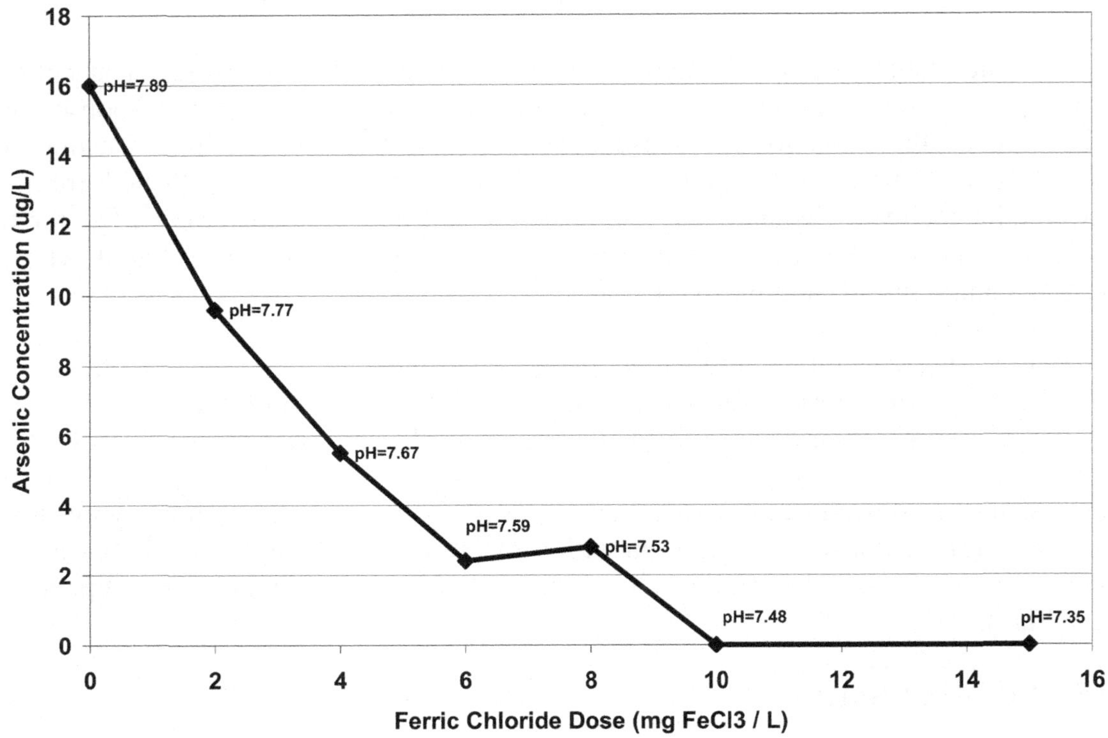

Figure 2.1 Impact of ferric chloride dose on finished water arsenic concentration (Run A, initial As = 16 μg/L)

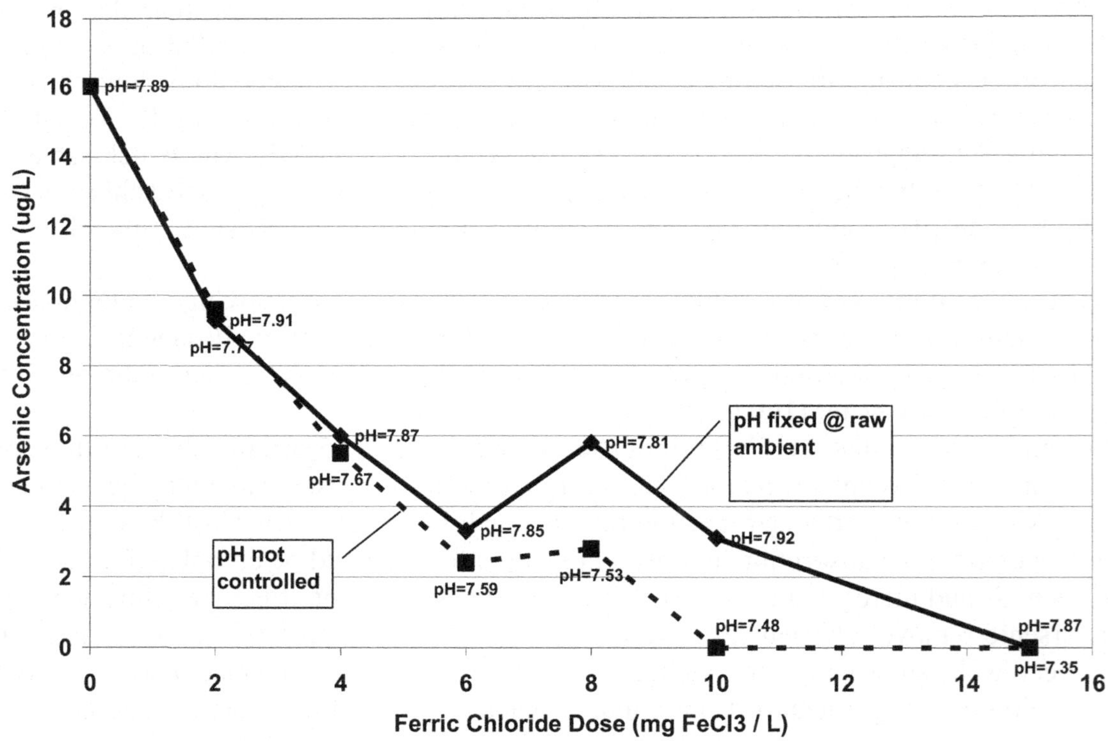

Figure 2.2 Comparison of arsenic removal at various coagulant dosages with and without pH control (initial As = 16 μg/L)

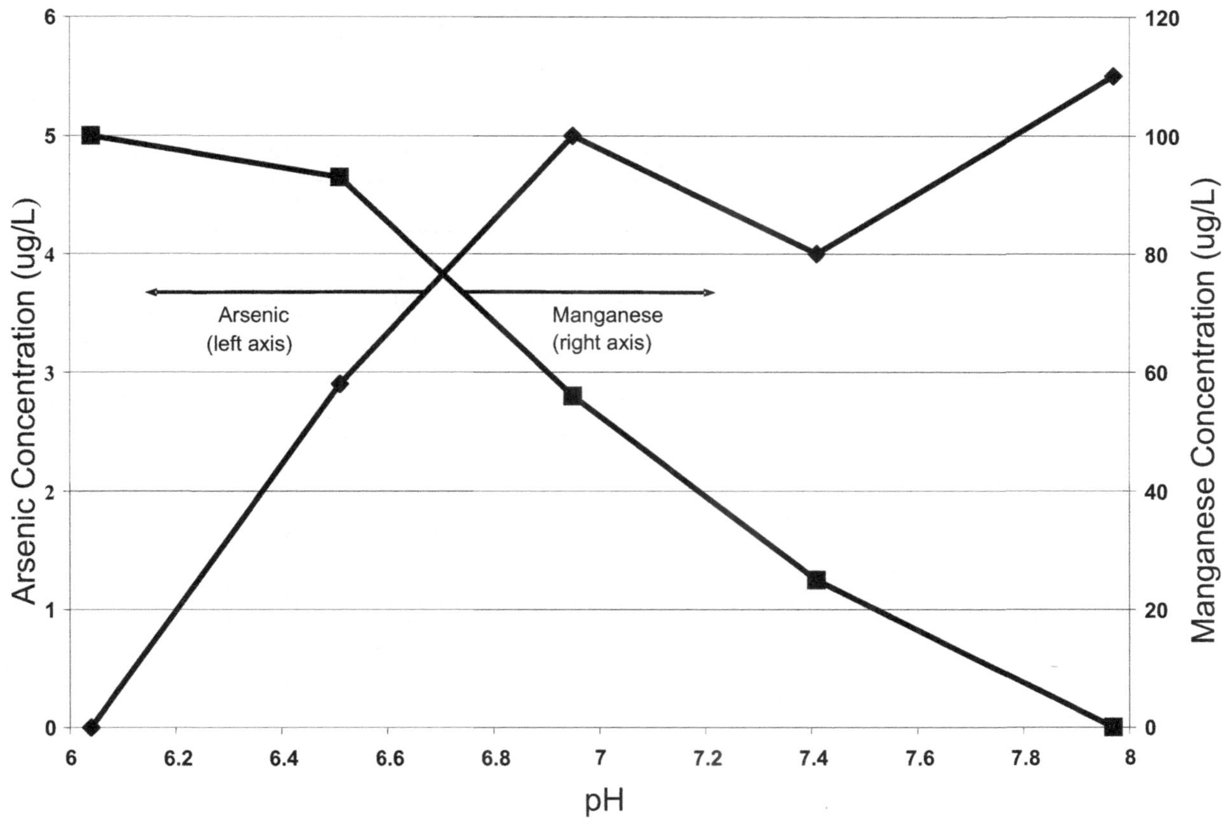

Figure 2.3 Impact of pH on arsenic and manganese removal with 4 mg/L ferric (initial As = 16 µg/L, Mn = 0.11 mg/L)

A jar test run was also conducted to investigate the addition of ferric chloride under controlled (or fixed) pH conditions (Run B). The pH selected as a target was the ambient raw water pH of 7.89. Figure 2.2 compares results from uncontrolled pH and fixed pH experiments. At low doses of ferric chloride, the pH is similar between the two runs and the finished arsenic concentration is also similar. At higher ferric chloride doses, the difference in pH was as much as 0.5 between the two runs with arsenic being removed more effectively at lower pH.

The impact of pH on arsenic removal was further investigated by fixing the ferric chloride dose at 4 mg $FeCl_3$/L and varying the pH (Run 3). Figure 2.3 illustrates that, for a given dose of ferric chloride, arsenic removal is superior at lower pH. Considering that a ferric dose between 4 and 6 mg/L could reduce arsenic levels below 6 µg/L with a minimal decrease in pH, adjusting pH either higher or lower will not provide justifiable benefit to the full-scale operation.

The impact of pH on manganese removal is also illustrated in Figure 2.3. Manganese concentrations were always below detection (<20 µg/L) in Runs A and B when the pH values were higher (Table 2.1). When the pH was intentionally adjusted lower in Run C, manganese concentrations in finished water significantly increased. Figure 2.3 illustrates the potential decrease in manganese removal efficiency as pH decreases. The bench testing is not expected to reliably predict manganese removal at full-scale, due to the catalytic behavior of the manganese dioxide coating on the full-scale filter media. However, the trend of decreased manganese removal at decreased pH is evident in the jar testing results and is consistent with the fact that manganese oxidation is slower at lower pH.

Table 2.1
Results from bench-scale testing

Sample	Ferric Chloride Dose (mg $FeCl_3$/L)	pH	Arsenic (μg/L)	Iron (μg/L)	Manganese (μg/L)
Run A: Variable Ferric Dose, Uncontrolled pH					
CU	0	7.89	16	ND	ND
A2	2	7.77	9.6	ND	ND
A4	4	7.67	5.5	ND	ND
A6	6	7.59	2.4	ND	ND
A8	8	7.53	2.8	ND	ND
A10	10	7.48	ND	ND	ND
A15	15	7.35	ND	ND	ND
Run B: Variable Ferric Dose, pH controlled to ambient					
CU	0	7.89	16	ND	ND
B2	2	7.91	9.3	ND	21
B4	4	7.87	6	ND	ND
B6	6	7.85	3.3	ND	ND
B8	8	7.81	5.8	ND	ND
B10	10	7.92	3.1	ND	ND
B15	15	7.87	ND	ND	ND
Run C: Sub-optimal Ferric Dose, Variable pH					
C8	4	7.97	5.5	ND	ND
C7.5	4	7.41	4	ND	25
C7.0	4	6.95	5	ND	56
C6.5	4	6.51	2.9	ND	93
C6.0	4	6.04	ND	ND	100

Notes: ND = Not detected at indicated Detection Limit for Reporting DLR (μg/L)

DLR	Parameter
2	Arsenic
100	Iron
20	Manganese

BENCH-SCALE TESTING CONCLUSIONS

- The addition of ferric chloride effectively controls finished water arsenic concentrations
- A dose of 4 to 6 mg $FeCl_3$/L was sufficient to achieve a finished water arsenic concentration target of 5 μg/L
- A dose of 4 to 6 mg $FeCl_3$/L will result in finished water pH of about 7.6
- A pH of 7.6 is not expected to significantly increase corrosion rates or metals release in the water system
- Ferric chloride is more effective at removing arsenic at lower pH
- Manganese removal will likely be less effective at lower pH
- Adjustment of pH appears to be unnecessary for this project

BENCH-SCALE TESTING RECOMMENDATIONS

- Proceed with full-scale demonstration using ferric chloride to remove arsenic
- Use a ferric chloride dose of 4 to 6 mg $FeCl_3$/L to achieve a target finished water arsenic concentration of 5 μg/L
- pH control is not required (either acid or base) and should not be installed at this time
- For the systems that require higher coagulant dosage, significant pH decrease may be expected. In order to avoid negative impact to manganese oxidation, the coagulant injection point should be downstream of the chlorination point and the contact time between these two chemical injections should long be enough for effective manganese oxidation.

CHAPTER 3
PILOT-SCALE TESTING

PILOT DESCRIPTION

Pilot-scale testing was conducted to evaluate filtration performance, to determine the relation between ferric chloride dose and filter run times and volumes, and to investigate higher hydraulic loading rates than were possible to achieve during full-scale testing. The pilot apparatus included a six-inch dual-media filter column with competent underdrain and equipment for headloss and flow monitoring (Figure 3.1). This pilot testing was conducted in parallel to the full-scale testing so that the scale-up issues and different hydraulic loading rates could be further investigated. Therefore, the feedwater flow to the pilot unit was the same as the full-scale filters that already received chlorine and ferric chloride addition.

The pilot scale column was 6 feet tall and 6 inches in diameter. Filtration media consisted of 18 inches of anthracite coal over 12 inches of silica sand (the same profile as the full-scale filters). New filtration media of similar specification to the full-scale media was loaded into the filtration column. The media was seasoned for effective manganese oxidation and removal by the method of Knocke et al. (1991). This method pre-conditions the media using a combination of potassium permanganate and chlorine. Various ferric dosages and filter loading rates were tested with this pilot system as shown in Table 3.1. The results from this pilot-scale study are compared to the full-scale test results to verify the suitability and accurateness of this treatment simulation method. Complete pilot testing results are located in Appendix B and summary charts are presented in this chapter for discussion.

Measured raw water quality (laboratory results) during pilot testing is presented in Table 3.2 for Wells 3 and 4. Flow weighted averaging was used to compute the expected blended raw water quality (filter influent), presented in Table 3.3. Table 3.4 presents general water chemistry of the well water during pilot and full-scale testing.

Figure 3.1 Pilot Filtration Column (6 inch diameter)

Table 3.1
Summary of pilot scale test results

Run Number	Start Date	***Filter Material***	Dose (mg/L FeCl3)	Loading Rate (gpm/ft2)	Break through Reached? (y/n)	Chart Location (Appendix B)
7	8/15/2002	dual media	8	5	y	B-3
8	8/19/2002	dual media	4	2.2	y	B-3
9	8/22/2002	dual media	4	8	y	B-4
10	8/26/2002	dual media	4	5	y	B-4
11	8/28/2002	dual media	6	5	y	B-5
12	9/3/2002	dual media	6	3	y	B-5

Table 3.2
Raw water quality during pilot scale tests (lab results)

Date/Time	Pilot Scale Run No.	Full Scale Run No.	WELL #3 Lab				WELL #4 Lab			
			Mn *(mg/L)* ***DL=0.02***	**Fe** *(mg/L)* ***DL=0.1***	**As** *(mg/L)* ***DL=0.002***	**Flow** *(gpm)*	**Mn** *(mg/L)* ***DL=0.02***	**Fe** *(mg/L)* ***DL=0.1***	**As** *(mg/L)* ***DL=0.002***	**Flow** *(gpm)*
8/1/2002 12:30	1	16	0.100	0.050	0.0088	490	0.071	0.050	0.016	477
8/5/2002 14:30	2	17	0.110	0.100	0.011	483	0.075	0.050	0.015	504
8/20/2002 9:30	8	24	0.098	0.050	0.0110	448	0.073	0.140	0.018	467
8/26/2002 13:00	10	26	0.110	0.130	0.0110	NM	0.077	0.050	0.016	NM
9/3/2002 8:25	12	29A	0.120	0.220	0.0085	483	0.082	0.050	0.014	510
Mean			0.108	0.110	0.010	476	0.076	0.068	0.016	490
Max			0.120	0.220	0.011	490	0.082	0.140	0.018	510
Min			0.098	0.050	0.009	448	0.071	0.050	0.014	467

Notes:
Values listed as 1/2 detection limit indicate parameter was not detected.
NM indicates parameter Not Measured.

Table 3.3
Calculated filter influent quality, pilot-scale tests

Date/Time	Pilot Scale Run No.	Filter Run No.	*Calculated Filter Influent* Mn *(mg/L)* *DL=0.02*	Fe *(mg/L)* *DL=0.1*	As *(mg/L)* *DL=0.002*	Flow *(gpm)*
8/1/2002 12:30	1	16	0.086	0.050	0.012	967
8/5/2002 14:30	2	17	0.092	0.074	0.013	987
8/20/2002 9:30	8	24	0.085	0.096	0.015	915
8/26/2002 13:00	10	26	0.093	0.090	0.014	963
9/3/2002 8:25	12	29A	0.100	0.133	0.011	993
	Mean		0.089	0.072	0.014	963
	Max		0.101	0.133	0.017	993
	Min		0.052	0.050	0.011	884

Notes:
Concentration is flow-weighted average using Well #3 and Well #4.
Flow is sum from Well #3 and Well #4.
Where flow parameter not measured, average well flow condition assumed for flow-weighted average

Table 3.4
General raw water quality during testing (6/17/02)

Parameter	Unit	Detection Limit	Well #3	Well #4
Arsenic III	*(mg/L)*	*DL=0.002*	0.0053	0.0058
Dissolved As	*(mg/L)*	*DL=0.002*	0.011	0.0140
Fluoride	*(mg/L)*	*DL=0.10*	0.34	0.32
PO_4	*(mg/L)*	*DL=0.02*	0.42	0.31
SiO_2	*(mg/L)*	*DL=0.5*	48	46
Total Alk	*(mg/L)*	*DL=1.0*	190	180
TDS	*(mg/L)*	*DL=1.0*	270	260
TOC	*(mg/L)*	*DL=0.5*	ND	ND
pH			7.8	7.9

PILOT TESTING RESULTS

Concentrations of iron, manganese, and arsenic in the pilot effluent were monitored and used as the indicators for filter breakthrough. Figure 3.2 illustrates arsenic, manganese, and iron concentrations during a typical filter run with 4 mg/L $FeCl_3$ operated at a hydraulic loading rate of 2.2 gpm/ft^2. Arsenic was reduced from 16 μg/L to below 7 μg/L through the filter run. The filter run time was determined by the breakthrough of iron, manganese, or arsenic, whichever came first. Iron broke through and reached the secondary MCL after 67 hours of filter operation. Arsenic concentrations increased slightly with iron concentrations at the end of the filter run. Manganese was not detected in filtered water throughout the run.

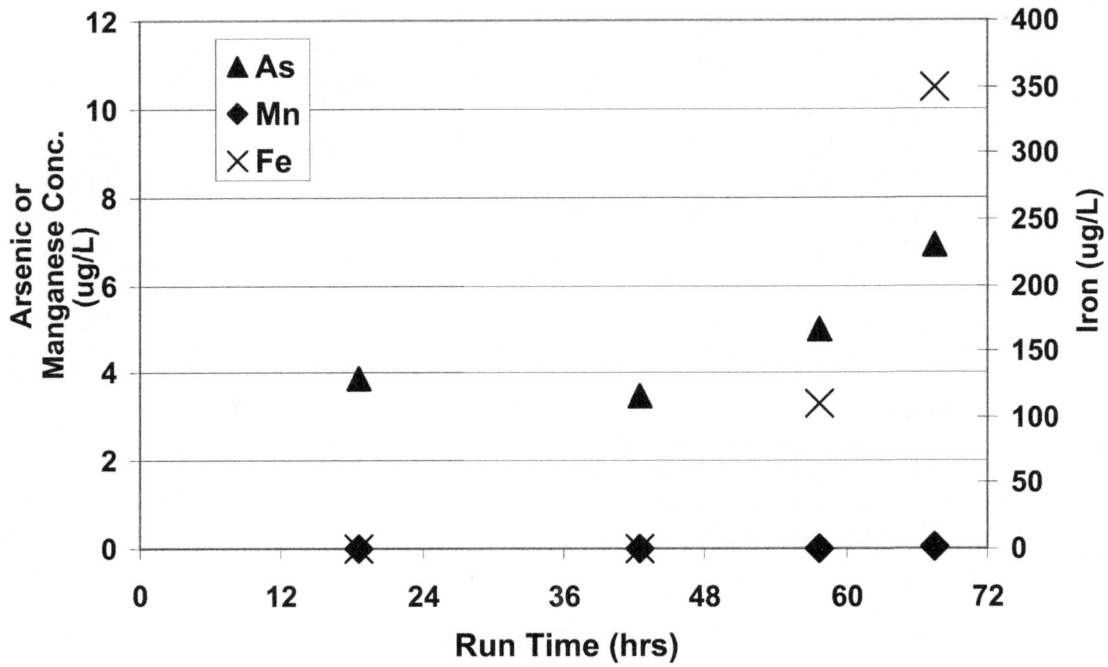

Figure 3.2 Pilot test run No.8 with a ferric chloride dosage of 4 mg/L and a loading rate of 2.2 gpm/ft^2

Figure 3.3 illustrates the impact of various filter loading rates and ferric chloride doses on filtered water arsenic concentrations. The figure presents all filtered water arsenic concentrations measured during pilot testing, as a function of dose and loading rate of the filter run. Arsenic concentrations plotted as zero indicate that arsenic was not detected with a 2 μg/L detection limit. At any given dose and loading rate, some variability is observed in the filtered water arsenic concentration due to analytical uncertainties and the stage in the filter run (i.e., filter ripening, maturity, or breakthrough). The results displayed in the figure do not suggest that filter loading rates significantly impact filtered water arsenic concentrations within the 2.2 through 8 gpm/ft^2 range examined. As expected, increasing ferric chloride doses result in lower filtered water arsenic concentrations.

Figure 3.4 summarizes the pilot-test results with dual filtration media. The figure illustrates the impact of filter hydraulic loading rate and ferric chloride dose on uniform filter run volume (UFRV). The UFRV expresses the total volume of water treated until breakthrough per square foot of filter surface. This normalized parameter is useful when comparing results from filters with different surface areas. Therefore, the concept of UFRV is used to evaluate and compare the effective filter production under various conditions. For example, a specific operating condition that produces an UFRV of 10,000 gal/ft^2 is considered more productive than an operating condition that produces an UFRV of 5,000 gal/ft^2.

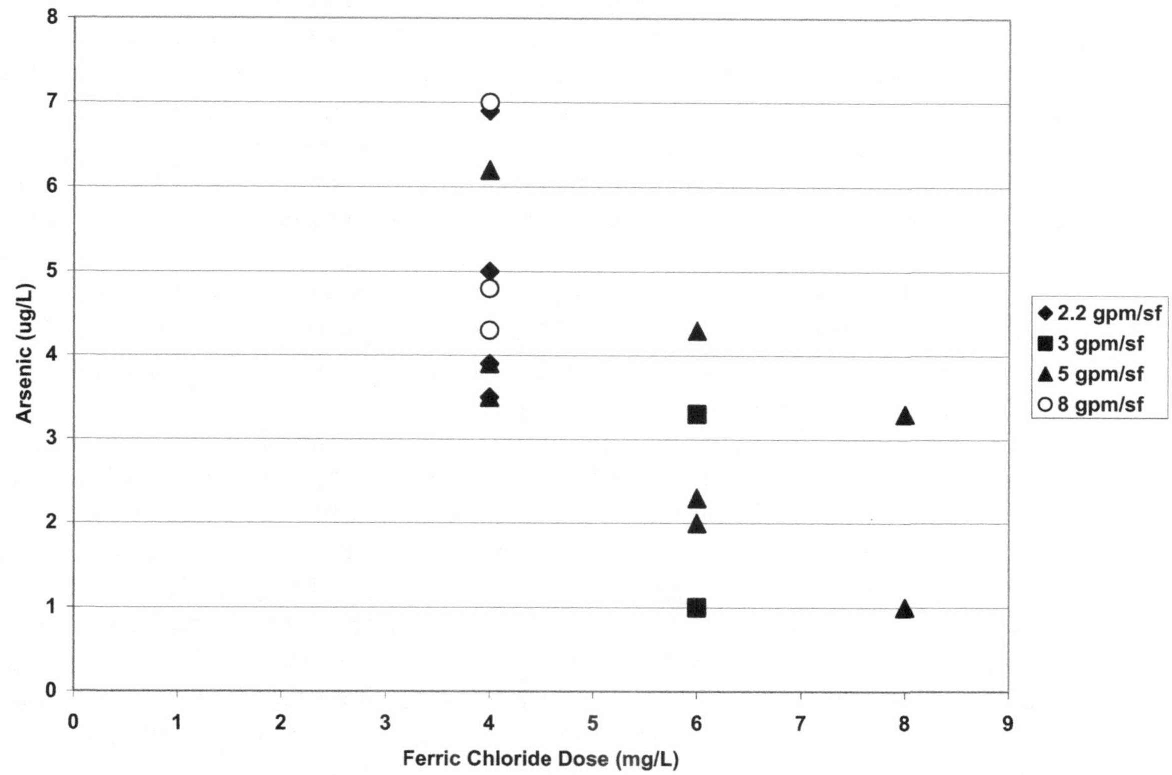

Figure 3.3 Impact of hydraulic loading rate and ferric chloride dose on filtered water arsenic concentration pilot-scale results

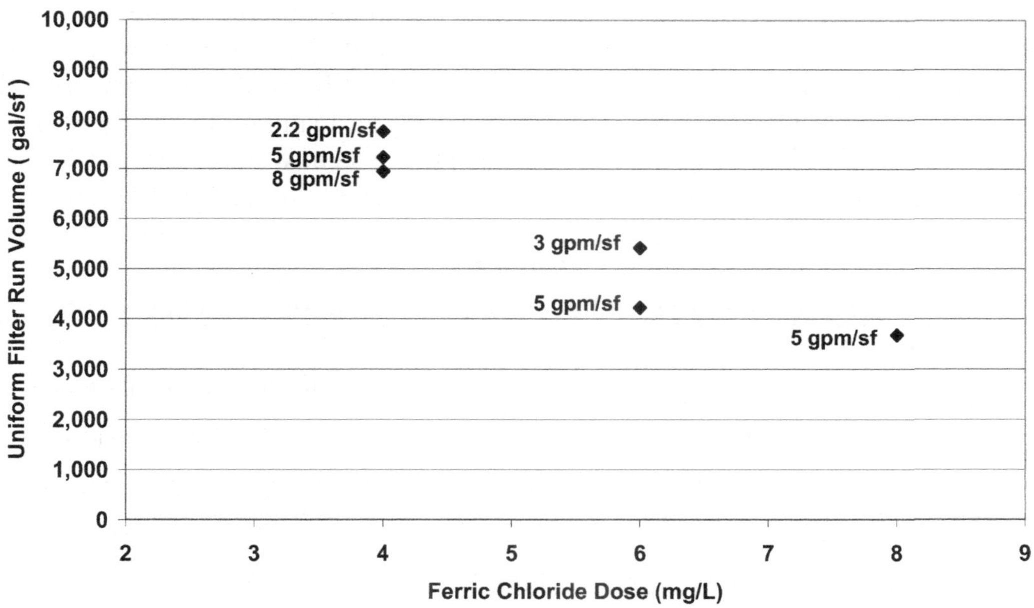

Figure 3.4 Impact of ferric dose on uniform filter run volume during pilot-scale tests

In the treatment facility planning process, UFRV can also be used to estimate the run time of a selected system. Assuming, for example, that the UFRV is the same (3,600 gal/ft^2) for the hydraulic loading rates of 3 gpm/ft^2 and 5 gpm/ft^2, to treat 1,000 gpm, 333 ft^2 of filter surface area would be required at 3 gpm/ft^2 or 200 ft^2 would be required at 5 gpm/ft^2. Each system would have the same water recovery (volume of product water divided by the volume of feed water) because the total quantity of water treated will be the same. The filter run time (the time between backwashes) at 3 gpm/ft^2 would be 20 hours and the filter run time at 5 gpm/ft^2 would be 12 hours.

Figure 3.4 shows that increasing ferric chloride dose decreases the UFRV. The UFRV decreased from 7,200 gal/ft^2 to 4,200 gal/ft^2 and 3,800 gal/ft^2 when the ferric chloride dose increased from 4 mg/L to 6 mg/L and 8 mg/L, respectively. This UFRV reduction is primarily due to the greater amount of solids in the feed water.

The figure also illustrates that the UFRV decreases slightly with increasing hydraulic loading rate. With a 4 mg/L ferric chloride dose, adjusting the hydraulic loading rate from 2.2 to 8 gpm/ft^2 reduced the UFRV from 7,700 gal/ft^2 to 6,900 gal/ft^2. This impact of loading rate to UFRV was more significant at higher ferric dosage. With a 6 mg/L ferric chloride dose, adjusting the loading rate from 3 gpm/ft^2 to 5 gpm/ft^2 reduced the UFRV from 5,500 gal/ft^2 to 4,200 gal/ft^2. The reduction of UFRV at higher loading rate could be attributed to the increased apparent flow velocity through the filter media, thereby increasing the shearing force exerted on the coagulant flocs attached on the media surface and limiting the amount of solids that can be retained by the filter. The impact of this increased shear force was less significant when the solid content in the influent was less (e.g., at 4 mg/L of ferric chloride), but was quite significant at a higher influent solid concentration (e.g., at 6 mg/L).

Table 3.5 presents an example of the impact of hydraulic loading rate on the design and operation of a filtration system. The table shows requirements to achieve 1,000 gpm water production under loading rate alternatives of 2.2, 5, and 8 gpm/ft^2. The filter surface area is inversely proportional to the loading rate so the filters sized at 2.2 gpm/ft^2 are much larger than those sized at 8 gpm/ft^2. Although the UFRV produced at each hydraulic loading rate was observed to be similar, the filter run time is much shorter for the 8 gpm/ft^2 loading rate. Although the total run volume is greater for the 2.2 gpm/ft^2 loading rate, so is the backwash volume because the filtration area is much greater. The overall result is an equivalent feedwater recovery for the different loading rates. However, the higher loading rate would result in capital cost savings because the filters and the backwash supply system would be smaller. Each option would generate an average backwash waste volume of about 23,000 gallons per day.

Table 3.5
Example impact of hydraulic loading rate on design and operation

Water production rate (gpm)	Loading rate (gpm/ft^2)	Filter surface area (ft^2)	UFRV (gal/ft^2)	Run time (hrs)	Run volume (gallons)	Backwash rate (gpm/ft^2)	Backwash duration (min)	Backwash volume (gallons)	Feedwater recovery (%)
1,000	2.2	455	7,757	57.7	3,526,115	12	10	54,545	98.5
1,000	5	200	7,233	25.2	1,446,522	12	10	24,000	98.4
1,000	8	125	6,945	14.8	868,089	12	10	15,000	98.3

PILOT TESTING CONCLUSIONS

- A ferric chloride dose of 4 mg/L appears sufficient to achieve a target filtered water arsenic concentration of 7 μg/L.
- Dosing ferric chloride does not appear to impact iron and manganese removal.
- Seasoned filtration media (18-inches coal over 12-inches sand) effectively removed manganese even with little oxidation time.
- Iron breakthrough controlled the filter run length.
- Increasing the ferric chloride dose decreases the filtered water arsenic concentration and decreases UFRVs.
- With 4 mg/L of ferric chloride, increasing hydraulic loading rate from 2.2 to 8 gpm/ft^2 did not impact filtered water arsenic concentrations and only slightly decreased UFRVs.

CHAPTER 4
FULL-SCALE TREATMENT SYSTEM DESIGN AND CONSTRUCTION

TREATMENT MODIFICATION

The SCWC Centralia Well Site was used for the full-scale demonstration of the coagulation/filtration technology. This chapter describes the modification of the existing manganese removal system to enhance arsenic removal. The operations and monitoring plan for the testing is also described.

As described in an earlier section, the Centralia Well Site consists of Well 3 and Well 4, producing a combined maximum capacity of 1,150 gpm. Wells typically operate in automatic mode with an operator onsite about one hour per day. Filter operation is controlled by a Modicon model 984-120 PLC. Filters are backwashed regularly according to a pre-set duration, but backwash can also be initiated manually. Backwash water is supplied from the distribution line leaving the site. Turbidity and chlorine residual are continuously monitored. Major modification of the treatment system included:

1. Installation of a ferric chloride feed system including a 500-gallon, mini-bulk storage tank, a calibration cylinder, and feed pump.
2. Installation of a static mixer with integrated feed port. The static mixer was installed in an underground vault to provide the capability of injecting ferric chloride into the combined flow stream (Figure 4.1 and Figure 4.2).
3. Installation of backwash holding tank. Temporary backwash waste holding tanks were installed to capture, settle, and decant backwash waste. After settling, supernatant was decanted for discharge to an adjacent stormwater channel under an existing NPDES permit. Sludge was retained in the bottom of the tanks until the completion of testing (Figure 4.3).
4. Instrumentation enhancements. Pressure transmitters were installed on filter inlet and outlet piping to determine headloss. Data loggers were configured to record flows and pressures. An automatic sampler was installed and programmed to collect filter effluent samples at specific times during the filter run. The modified process diagram is illustrated in Figure 4.4.

Figure 4.1 Example of a static mixer in a flanged pipe spool
Source: Komax Systems, Wilimington, CA

Figure 4.2 Static mixer (8 inch) with integrated feed port installed in underground vault

Figure 4.3 Temporary backwash waste holding tanks

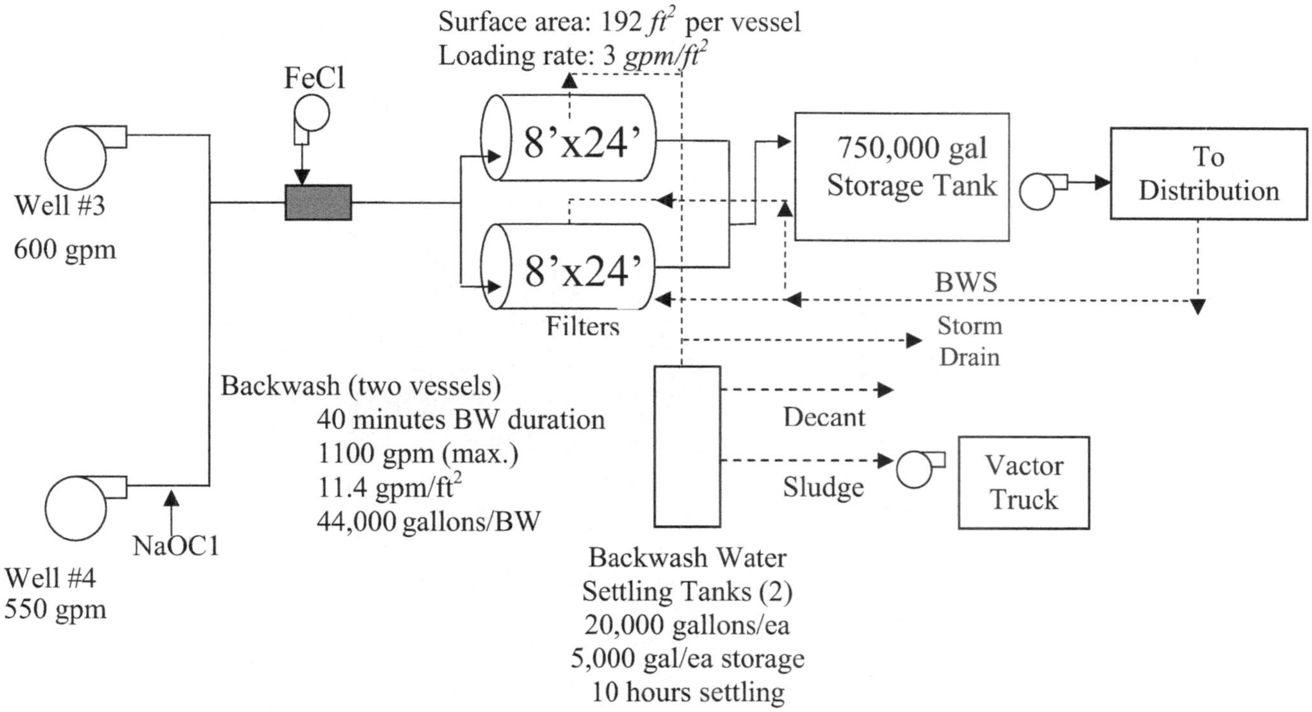

Figure 4.4 Centralia plant site after modification

CALCULATION OF FERRIC CHLORIDE CONSUMPTION

Ferric chloride was supplied in a 500-gallon, mini-bulk system (Figure 4.5). The ferric chloride solution consisted of about 43 percent $FeCl_3$ by weight with a specific gravity of 1.47, containing 632 mg of active $FeCl_3$ per mL of solution as follows:

$$\frac{1.47g\text{ soln}}{mL\text{ soln}} \times \frac{0.43g\ FeCl_3}{g\text{ soln}} \times \frac{1{,}000mg}{g} = 632\text{ mg FeCl}_3\text{/mL soln}$$

Ferric chloride was fed neat (undiluted) to the process stream. The maximum process water flow rate was anticipated to be 1,150 gpm with both Well Pumps 3 and 4 running. A typical ferric chloride dose of 8 mg $FeCl_3$/L would require a neat ferric chloride feed rate of 0.87 gallons per hour as follows:

$$\frac{1{,}150\,gal\ water}{\text{min}} x \frac{3.785L}{gal} x \frac{8mg\ FeCl_3}{L\ water} = 34{,}822\text{ mg FeCl}_3\text{/min}$$

$$\frac{34{,}822mg\ FeCl_3}{\text{min}} \times \frac{mL\text{ soln}}{632\,mg\ FeCl_3} = 55.1\text{ mL ferric soln/min}$$

$$\frac{55.1\,mL\text{ soln}}{\text{min}} \times \frac{60\,\text{min}}{hr} \times \frac{L}{1{,}000mL} \times \frac{gal}{3.785L} = 0.87 \text{ gallon ferric soln/hour}$$

or 21 gallons/day average use

Figure 4.5 Ferric chloride mini-bulk tank (500 gal) with secondary containment

CHEMICAL FEED SYSTEM

Ferric chloride was delivered from the mini-bulk container to the static mixer in neat form using a solenoid actuated diaphragm pump (Figure 4.6). The pump had variable speed control capability and adjustable stroke with a combined turndown ratio of 100:1. Pump speed was set manually to deliver the desired output. The pump was located in the chlorine building and was switched with the operation of Well Pump 3 through an existing 120-volt outlet on the west wall of the chlorine building.

Based on the need to deliver a maximum of 15 mg/L ferric chloride dose, the diaphragm pump had a capacity of 40 gallons/day to accommodate a maximum flow of 39 gallons/day as follows:

$$\frac{1{,}150\,gal}{\text{min}} \times \frac{15\,mg\,FeCl_3}{L} \times \frac{L}{1{,}000\,mL} \times \frac{mL\text{ soln}}{632\ mg\ FeCl_3} \times \frac{60\,\text{min}}{hr} \times \frac{24hr}{day} = 39 \text{ gallons/day}$$

The dosage rate was controlled as described in the Operating and Monitoring Plan.

Figure 4.6 Ferric chloride feed pump and calibration cylinder

The chemical feed pump suction line was a polyethylene flexible hose, which connected with female quick-disconnect connections at each end to a male quick-disconnect on the discharge valve of the mini-bulk container, and to piping at the outside of the wall of the chlorine building (Figure 4.7). The suction line was routed through the wall of the chlorine building and included a 2 inch diameter by 1 foot, 6 inch plastic calibration column and valving used to calibrate the feed pump. The calibration column was placed on the floor of the chlorine building to allow it to be charged with gravity head.

A white polyethylene ¼-inch tubing discharge line was inserted in 1.5-inch PVC electrical conduit with radius bends to connect the ferric metering pump to the injection port of the static mixer. The conduit was buried a minimum of 1 foot below grade in a trench.

A spare empty 1.5-inch conduit was also provided. The discharge line penetrated the static mixer vault wall and connected to the injection port. Standard injection fittings were provided as well as check valve to facilitate disassembling the discharge piping from the injection port without draining the static mixer.

DESIGN OF STATIC MIXER AND ACCESS VAULT

A PVC two-element static mixer was installed in the existing 8-inch raw water line after lines from Well Pumps 3 and 4 are combined following chlorination (Figure 4.8). Butterfly valves were installed at each end of the mixer so it could be isolated. The mixer and valves were enclosed in a buried FRP Armorcast© vault with a galvanized steel top frame and polymer torsion assist covers.

The vault was six feet inside depth to accommodate the existing pipe, which has a four foot depth to the top of the pipe. The vault contained a sump to a gravel base below the concrete support pad to drain or pump any water released in the vault. The static mixer was made of PVC, 8 inches in diameter and 34 inches long, and included an integrated 2 inch threaded additive injection port.

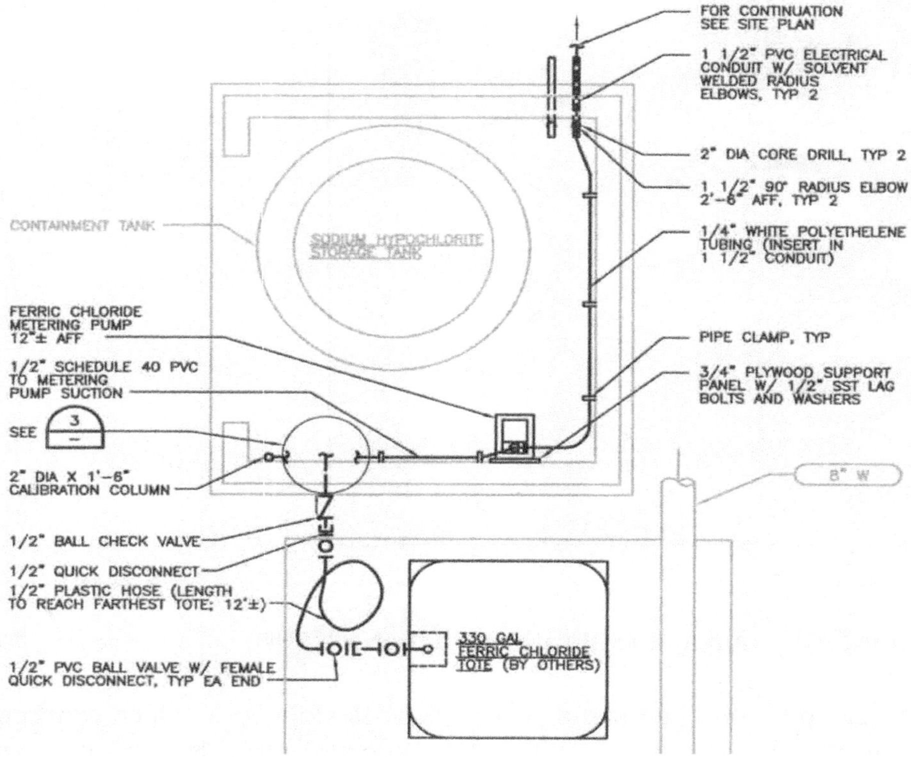

Figure 4.7 Ferric chloride feed system (tote containers replaced with mini-bulk tank)

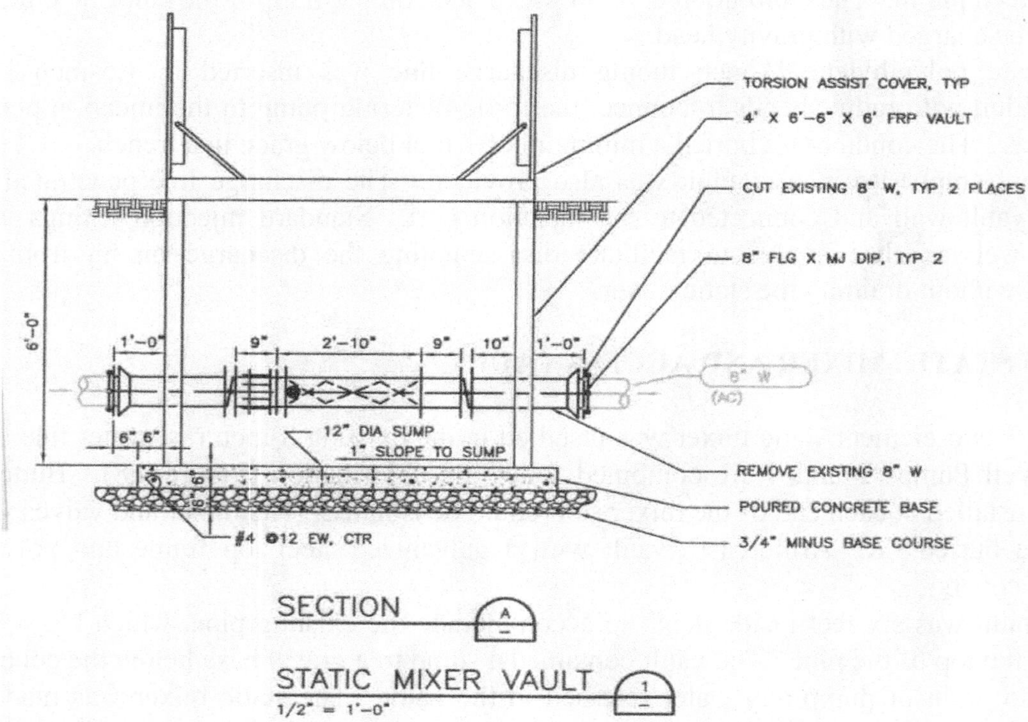

Figure 4.8 Vault with static mixer

RESIDUALS HANDLING

The quantity of solids generated by the existing manganese removal system will increase due to the addition of ferric chloride coagulant. Solids are trapped in the filter during normal filtration process and removed from the filter during filter backwash. Filter backwash waste flowed to two backwash holding tanks (40,000 gallons total storage). After settling, supernatant was discharged to the adjacent stormwater channel through an electrically activated 4-inch ball valve working off a discrete contact from the filter PLC. Consolidated residuals (about 10,000 gallons) were retained in the bottom of the tank until the end of the project.

Solids Generation

Solids are produced by the existing process through the oxidation of iron with precipitation of ferric hydroxide [$Fe(OH)_3$] and the oxidation of manganese with the precipitation of manganese dioxide [MnO_2]. Solids generation increased due to the addition of ferric chloride that precipitates to form ferric hydroxide. Ferric chloride could contain a significant amount of manganese. The utility could obtain the quantity of manganese in ferric chloride from their chemical supplier and incorporate that into calculations regarding sludge quality and quantity. The calculations below assume that the concentration of manganese in ferric chloride is not significant.

The average dose of ferric chloride used in this study is expected to be in the range of 8 mg $FeCl_3$/L. The native iron in the water is less than 0.1 mg Fe/L. The native manganese in the water is in the range of 0.09 mg Mn/L.

Reactions producing solids:

Dissociation of ferric chloride

$FeCl_3 \rightarrow Fe^{3+} + 3Cl^-$

Precipitation of ferric hydroxide

$Fe^{3+} + 3H_2O \rightarrow Fe^{3+}(OH^-)_3\ (s) + 3H^+$

Precipitation of manganese dioxide

$Mn^{4+} + 2H_2O \rightarrow Mn^{4+}O_2\ (s) + 4H^+$

Solids Production:

Dosed Iron

$$\frac{8mgFeCl_3}{L} \times \frac{1mol}{162.4g} \times \frac{1molFe}{1molFeCl_3} \times \frac{1molFe(OH)_3}{1molFe}$$

$$\times \frac{106.9gFe(OH)_3}{molFe(OH)_3} \times 1{,}150gpm \times \frac{1MGD}{694.5gpm} \times \frac{8.34L-lb}{MG-mg}$$

= 72.7 lb $Fe(OH)_3$ solids/day

Native Iron

$$\frac{0.1 mgFe}{L}\times\frac{molFe}{55.9gFe}\times\frac{molFe(OH)_3}{molFe}\times\frac{106.9gFe(OH)_3}{molFe(OH)_3}\times 1{,}150gpm\times\frac{1MGD}{694.5gpm}\times\frac{8.34L-lb}{MG-mg}$$

= 2.6 lb $Fe(OH)_3$ solids/day

Native Manganese

$$\frac{0.09mgMn}{L}\times\frac{molMn}{54.9g}\times\frac{molMnO_2}{molMn}\times\frac{86.9gMnO_2}{molMnO_2}\times 1{,}150gpm\times\frac{1MGD}{694.5gpm}\times\frac{8.34L-lb}{MG-mg}$$

= 2.0 lb MnO_2 solids/L

Total solids production were calculated to be 77 lb/day.

$72.7lb/day + 2.6lb/day + 2.0lb/day$ = 77.3 lb/day solids production

An alternative method for calculating the solids production was presented by Awwa (1998). The equation for calculating solids production is as follows:

SS = Q(2.9 Fe + SS + P)(8.34 lb/MG/mg/L)

where:

S =	solids produced, lbs dry solids/day
Q =	water production, million gallons
SS =	suspended solids concentration, mg/L
Fe =	ferric chloride dosage, mg/L (dosage as Fe)
P =	polymer dosage, mg/L

Assuming negligible suspended solids and no polymer addition, the solids produced by coagulant addition was determined to be 110.2 lbs/day.

$$2.9\times\frac{2.75mgFe\ added}{L}\times 1{,}150gpm\times\frac{1\ MGD}{694.5\ gpm}\times\frac{8.34L-lb}{MG-mg}$$

=110.2 lbs/day solids production

This method gives a greater sludge production and a more conservative design criterion. Therefore, this value was used for design purposes.

Backwash Waste Generation

The total backwash waste generated was in the range of 30,000 gallons per backwash cycle. This backwash waste was not expected to meet the existing NPDES permit. Therefore, it was contained, settled, and decanted prior to discharge to the stormwater channel. The backwash sequence is shown in Table 4.1.

Table 4.1
Backwash sequence

Item	Duration (min)	Rate (gpm)	Total Volume (gallons)
Surface Wash Filter 1, Cell A	1		
Backwash Filter 1, Cell A	6	1,000	6,000
Surface Wash Filter 1, Cell B	1		
Backwash Filter 1, Cell B	6	1,000	6,000
Surface Wash Filter 2, Cell A	1		
Backwash Filter 2, Cell A	6	1,000	6,000
Surface Wash Filter 2, Cell B	1		
Backwash Filter 2, Cell B	6	1,000	6,000
Filter to Waste	3	1,150	3,450
Total			27,450

FULL-SCALE SYSTEM OPERATIONS AND MONITORING

An operations and monitoring plan was developed to guide the project. The monitoring program included:

- Tasks conducted twice per day during filter runs of three days or less, and once per day for longer filter runs
  - Field analysis of iron and manganese (influent and effluent) (Figure 4.9)
  - Lab analysis of arsenic, iron, and manganese (effluent only)
  - Field analysis of chlorine (effluent only)
- Tasks conducted once per day
  - Parameters recorded with date and clock time
    - Influent and effluent pressures
    - Flow from both wells
    - Turbidity
    - Filter time in operation
  - Activities conducted once per filter run
  - Lab analysis of arsenic, iron, and manganese (influent)
  - Verify chemical feed pump output
  - Record date and time for start of filter run
  - Record date and time for filter-to-waste
  - Record date and time for start of backwash
  - Use samples stored in autosampler to characterize breakthrough (3 samples for lab analysis)

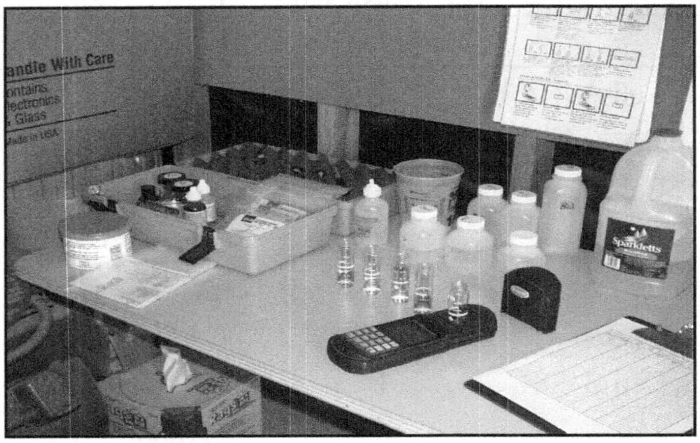

Figure 4.9 Bench workspace for field analysis of iron and manganese

The dosing of ferric chloride began at 2 mg/L and was increased by 2 mg/L in each successive filter run. The increasing of the ferric chloride dose ended when filter run lengths declined to 15 hours or less, which is the shortest run length that SCWC would like to operate at due to operational constraints. By increasing the dose in successive filter runs in this manner, the operational characteristics of the filtration system were rapidly defined while maintaining control of filtered water quality.

Once the operational characteristics of the filters had been observed over the full range of ferric chloride doses, selected ferric chloride doses were repeated to confirm the results. The repeated doses were 4 and 6 mg/L as $FeCl_3$.

CONSTRUCTION AND STARTUP

Equipment and materials required for the construction and startup of the project were shipped directly from equipment suppliers to the testing site and installed. Following construction, HDR personnel visited the site to assist the startup of the treatment process and provide assistance to field engineers and operators of SCWC.

EQUIPMENT AND MATERIALS

Equipment and materials required for full-scale testing were constructed and installed during the period of April 22 to May 3, 2002. Appendix C provides a photo log of the test site. Equipment installed included:

- Mini-bulk tank for ferric chloride (500 gallons)
- Chemical feed calibration cylinder
- Chemical feed pump and discharge piping
- Static mixer in vault
- Backwash holding tanks, piping, and automated decant valve
- Dechlorination port for backwash water

- Pressure transmitters
- Data loggers (2) monitoring differential pressure, flow, and turbidity
- Autosampler

ONSITE ASSISTANCE

HDR provided onsite assistance during full-scale filter run start-up and associated analysis. Onsite coordination consisted of the following:

- Monitoring methods
- Metering pump installation
- Chemical feed calibration cylinder
- Installation of auto sampler and data loggers (Figure 4.10)
- Actuated valve for backwash decant

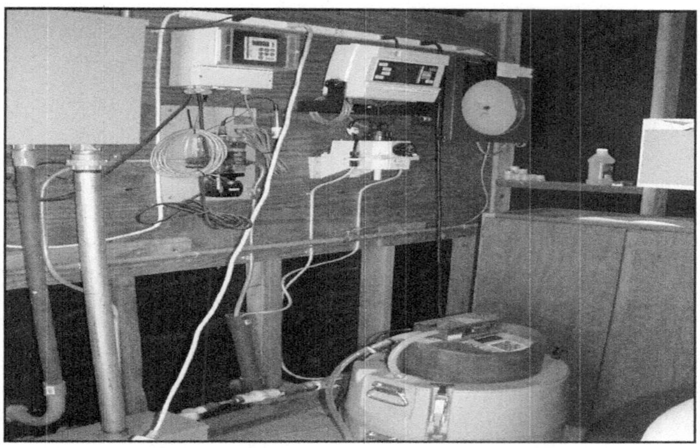

Figure 4.10 Centralia Plant Instrumentation with auto sampler in foreground

CHAPTER 5
FULL-SCALE TESTING RESULTS AND DISCUSSION

The full-scale testing evaluated the impact of ferric chloride dose on filtered water arsenic concentrations, filter run times, and UFRV. Process design criteria and testing procedures are detailed in Chapter 4, Full-Scale Treatment System Process Design and Construction. Each filter run was operated under a constant ferric chloride dose and hydraulic loading rate throughout the run.

Table 5.1 provides a comprehensive list of the filter runs performed, dates, ferric chloride dosages, and whether the filter was run until breakthrough of iron. Comprehensive results for all filter runs are also included in Appendix D, in the form of charts illustrating filtered water arsenic concentration as a function of time during the filter run. Because the comprehensive data is provided in the appendix, this chapter provides a summary review of that data using specific filter run results as examples to illustrate trends and conclusions.

RAW WATER QUALITY

Measured raw water quality (laboratory results) during full-scale testing is presented in Table 5.2 for Wells 3 and 4. Flow weighted averaging was used to compute the expected blended water quality (filter influent), presented in Table 5.3. Table 3.4 presents general water chemistry of the well water during pilot and full-scale testing.

SIMULTANEOUS REMOVAL OF ARSENIC, IRON, AND MANGANESE

As demonstrated in both bench and pilot testing (Chapters 2 and 3) the process of chlorination, ferric chloride addition, and filtration can effectively remove arsenic, iron, and manganese simultaneously. The effective removal of these inorganic contaminants in a full-scale pressure filtration system can be seen in all of the charts in Appendix D, where iron and manganese were typically removed to below detection limits and arsenic removal depended on the operating ferric chloride dose. The full-scale system was operated with a relatively constant hydraulic loading rate of 2.2 gpm/ft^2 and each run was performed with a constant ferric chloride dose. As observed in the charts of Appendix D, iron typically broke through the filter prior to manganese or arsenic breakthrough, although there would presumably be some arsenic associated with the iron breaking through the filter. Nonetheless, the quantity of arsenic associated with this passing iron particles should be proportional to the percentage of iron passing and should be relatively insignificant. Headloss was monitored during the filter run. Headlosses never increased more than 2 psi over the headloss at the start of the filter run.

Figure 5.1 provides an example filter run illustrating these typical observations. Filter Run No. 7 illustrated in the figure was operated with a ferric chloride dose of 4 mg/L as ferric chloride. Iron and manganese were removed to below detection throughout the run until hour 56 when iron broke through the filter. Filtered water arsenic concentrations were higher at the beginning of the run but then stabilized during the run within a range of 4 to 6 µg/L. The higher arsenic concentrations at the beginning of the run result from using backwash water that contained arsenic in the range of 14 µg/L, as will be explained later in this chapter.

Table 5.1
Summary of full scale test runs

Run Number	Start Date	Dose (mg/L FeCl3)	Break-through Reached? (y/n)	Chart Location (Appendix D)
1	5/2/2002	2	n	D-3
2	5/6/2002	2	n	D-3
3	6/10/2002	2	y	D-4
4	6/17/2002	2	n	D-4
5	6/19/2002	2	n	D-5
6	6/21/2002	2	n	D-5
7	6/24/2002	4	y	D-6
9	7/8/2002	6	y	D-6
10	7/10/2002	8	y	D-7
11	7/16/2002	10	y	D-7
12	7/17/2002	10	y	D-8
14	7/18/2002	14	y	D-8
15	7/19/2002	2	n	D-9
16	8/1/2002	4	n	D-9
17	8/5/2002	4	n	D-10
18	8/6/2002	4	n	D-10
19	8/7/2002	6	n	D-11
20	8/8/2002	6	n	D-11
21	8/12/2002	6	y	D-12
22	8/14/2002	8	y	D-12
23	8/15/2002	8	n	D-13
24	8/19/2002	4	n	D-13
25	8/21/2002	4	n	D-14
26	8/26/2002	4	n	D-14
27	8/27/2002	6	n	D-15
28	8/28/2002	6	n	D-15
29A	9/3/2002	6	n	D-15
30B	9/13/2002	4	n	D-16
30	10/1/2002	6	n	D-16
31	10/3/2002	6	n	D-16

Table 5.2
Raw water quality during full scale tests (lab results)

Date/Time	Pilot Scale Run No.	Full Scale Run No.	WELL #3 Lab Mn *(mg/L) DL=0.02*	WELL #3 Lab Fe *(mg/L) DL=0.1*	WELL #3 Lab As *(mg/L) DL=0.002*	WELL #3 Lab Flow *(gpm)*	WELL #4 Lab Mn *(mg/L) DL=0.02*	WELL #4 Lab Fe *(mg/L) DL=0.1*	WELL #4 Lab As *(mg/L) DL=0.002*	WELL #4 Lab Flow *(gpm)*
5/7/2002 8:30		2	0.120	0.130	0.0110	NM	0.081	0.050	0.018	NM
6/10/2002 8:40		3	0.095	0.050	0.0130	483	0.060	0.050	0.021	503
6/17/2002 7:00		4	0.110	0.120	0.0110	449	0.079	0.050	0.018	435
6/24/2002 12:00		7	0.120	0.110	0.0130	481	0.080	0.050	0.018	482
7/8/2002 11:35		9	0.092	0.100	0.0190	477	0.087	0.120	0.015	515
7/17/2002 7:40		11	0.100	0.050	0.0110	477	0.070	0.050	0.017	477
7/22/2002 7:00		15	0.098	0.050	0.0110	477	0.070	0.050	0.018	477
8/1/2002 12:30	1	16	0.100	0.050	0.0088	490	0.071	0.050	0.016	477
8/5/2002 14:30	2	17	0.110	0.100	0.011	483	0.075	0.050	0.015	504
8/12/2002 17:00		21	0.100	0.050	0.0098	483	0.072	0.050	0.016	510
8/20/2002 9:30	8	24	0.098	0.050	0.0110	448	0.073	0.140	0.018	467
8/26/2002 13:00	10	26	0.110	0.130	0.0110	NM	0.077	0.050	0.016	NM
9/3/2002 8:25	12	29A	0.120	0.220	0.0085	483	0.082	0.050	0.014	510
9/9/2002 7:30		29A	0.120	0.050	0.0083	NM	0.082	0.050	0.013	NM
9/17/2002 8:30		29B	0.110	0.050	0.0097	NM	0.078	0.050	0.016	NM
10/2/2002 14:45		30	0.094	0.050	0.0110	NM	0.010	0.050	0.014	NM
Mean			0.106	0.085	0.011	476	0.072	0.060	0.016	487
Max			0.120	0.220	0.019	490	0.087	0.140	0.021	515
Min			0.092	0.050	0.008	448	0.010	0.050	0.013	435

Notes:
Values listed as 1/2 detection limit indicate parameter was not detected.
NM indicates parameter Not Measured.

Table 5.3
Calculated filter influent quality, full-scale tests

Date/Time	Pilot Scale Run No.	Filter Run No.	Calculated Filter Influent			
			Mn *(mg/L)* *DL=0.02*	**Fe** *(mg/L)* *DL=0.1*	**As** *(mg/L)* *DL=0.002*	**Flow** *(gpm)*
5/7/2002 8:30		2	0.100	0.090	0.015	963
6/10/2002 8:40		3	0.077	0.050	0.017	986
6/17/2002 7:00		4	0.095	0.086	0.014	884
6/24/2002 12:00		7	0.100	0.080	0.016	963
7/8/2002 11:35		9	0.089	0.110	0.017	992
7/17/2002 7:40		11	0.085	0.050	0.014	954
7/22/2002 7:00		15	0.084	0.050	0.015	954
8/1/2002 12:30	1	16	0.086	0.050	0.012	967
8/5/2002 14:30	2	17	0.092	0.074	0.013	987
8/12/2002 17:00		21	0.086	0.050	0.013	993
8/20/2002 9:30	8	24	0.085	0.096	0.015	915
8/26/2002 13:00	10	26	0.093	0.090	0.014	963
9/3/2002 8:25	12	29A	0.100	0.133	0.011	993
9/9/2002 7:30		29A	0.101	0.050	0.011	963
9/17/2002 8:30		29B	0.094	0.050	0.013	963
10/2/2002 14:45		30	0.052	0.050	0.013	963
	Mean		0.089	0.072	0.014	963
	Max		0.101	0.133	0.017	993
	Min		0.052	0.050	0.011	884

Notes:
Concentration is flow-weighted average using Well #3 and Well #4.
Flow is sum from Well #3 and Well #4.
Where flow parameter not measured, average well flow condition assumed for flow-weighted average

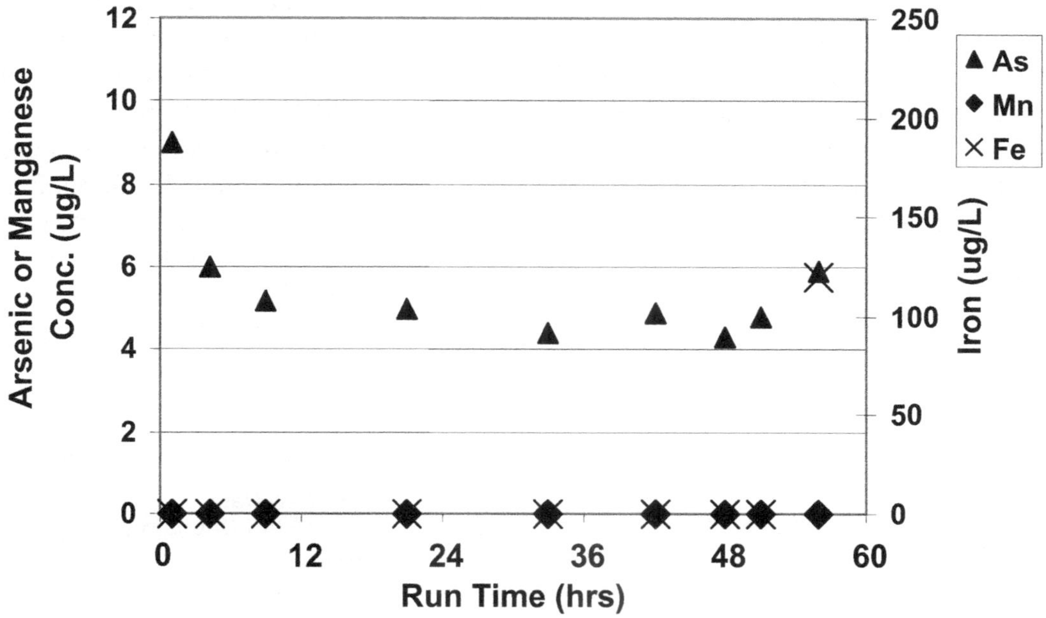

Figure 5.1 Removal of arsenic, iron, and manganese under 4 mg/L $FeCl_3$ dose (influent As: 14 µg/L)

Because filtered water iron and manganese concentrations at Centralia site were generally not detectable during any filter run in this study, it does not appear that the ferric chloride dose had any meaningful impact on iron and manganese removal. Therefore, simultaneous removal of arsenic, iron, and manganese is possible with the treatment process investigated. Utilities that are currently removing iron and manganese by oxidation and filtration have the opportunity to remove arsenic as well with little impact to filtered water iron and manganese concentrations. As discussed in the next section and later in this chapter, ferric chloride dose will impact filtered water arsenic concentrations and filter run times.

IMPACT OF FERRIC CHLORIDE DOSE ON ARSENIC CONCENTRATION

Although the dose of ferric chloride did not impact the filtered water iron and manganese concentrations, it did impact filtered water arsenic concentrations. Figure 5.2 illustrates the impact of ferric chloride dose on the observed filtered water arsenic concentrations in a summary of all full-scale testing performed. Only results from filter run times greater than five hours were used in creating Figure 5.2. Greater ferric chloride doses result in lower filtered water arsenic concentrations. Arsenic removal appears to be reasonably linear for ferric chloride doses in the range of 0 to 4 mg/L. Doses from 8 to 14 mg/L showed diminishing returns in removing additional arsenic and resulted in filtered water arsenic concentrations near the detection limit (Figure 5.2).

The amount of arsenic removed per mass of ferric chloride added is similar for the 2 and 4 mg/L doses. For doses greater than 4 mg/L, filtered water arsenic concentrations are in the range of the 2 μg/L analytical detection limit, introducing additional uncertainties and preventing drawing clear conclusions about the mass of arsenic removed per mass of ferric chloride dosed. Table 5.4 lists a statistical summary of results and presents calculations related to the amount of arsenic that was removed per unit mass of ferric chloride added. Median filtered water arsenic concentrations are based on all full-scale filter runs performed and include values observed five hours or more into the filter run. The percent arsenic removed is similar for doses 2 and 4 with each 2 mg/L achieving approximately 30 percent removal of the original arsenic mass.

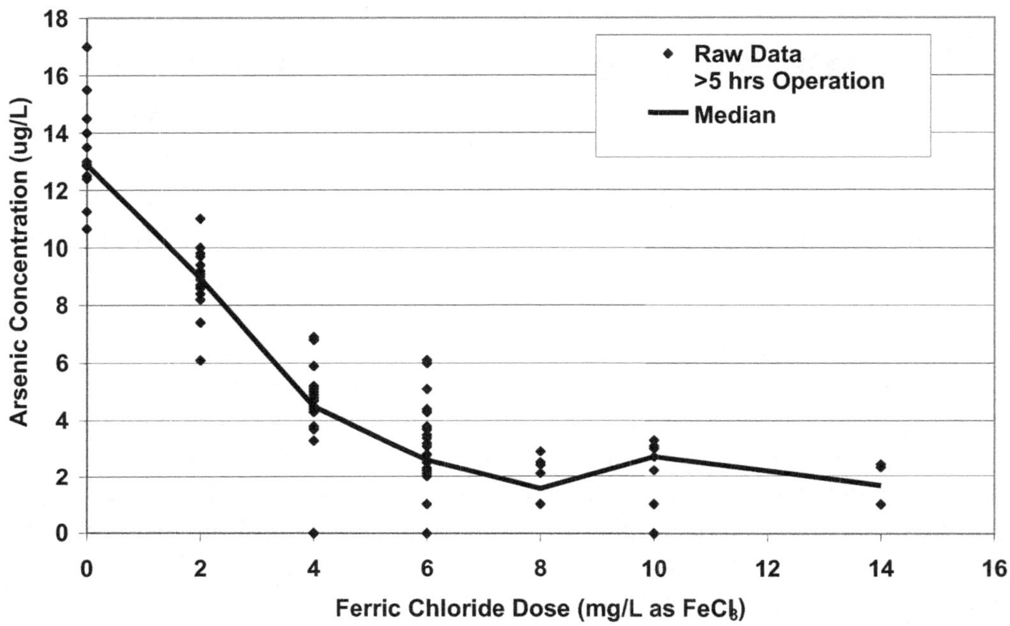

Figure 5.2 Impact of ferric chloride on filtered water arsenic concentration on full-scale results (raw water arsenic concentration represented by dose of 0)

Table 5.4
Median filtered water arsenic concentrations and calculation of removal as a function of ferric chloride dose

	Ferric Chloride Dose (mg/L as $FeCl_3$)						
	0	2	4	6	8	10	14
Median arsenic concentration (μg/L)	12.9	8.95	4.5	2.6	1.55	2.7	1.65
Fraction arsenic removed [(Co-C)/Co]	N/A	0.31	0.65	0.8	0.88	0.79	0.872
Incremental iron added (mg/L as $FeCl_3$)	0	2	2	2	2	2	4
Incremental iron added (mg/L as Fe)	0	0.69	0.69	0.69	0.69	0.69	1.377
Incremental arsenic removed (μg/L)	N/A	3.95	4.45	1.9	1.05	NC	NC
Total arsenic removed (μg/L)	N/A	3.95	8.4	10.3	11.35	10.2	11.25
Arsenic (μg/L) : iron (mgL as $FeCl_3$)	N/A	1.98	2.10	1.72	1.42	1.02	0.80
Arsenic (μg/L): iron (mgL as Fe)	N/A	5.74	6.18	5.03	4.18	3.0	2.35

Notes: Fe MW = 55.9, Cl MW = 35.5, $FeCl_3$ = 0.34
Where arsenic not detected, half detection limit used to compute median concentration.
NC: not computed because of limits of detection

Adding ferric chloride or increasing the ferric chloride dose decreases the filter run times and volume of water treated for a given hydraulic loading rate due to the additional solids added. Filter runs that were performed until breakthrough occurred are noted in Table 5.1. Figure 5.3 presents results from Filter Run No. 9, showing typical observations during breakthrough. The end of the filter run is marked by an increase in the filtered water iron concentration, which is determined as the breakthrough point of a run. Arsenic may increase slightly corresponding with the iron breakthrough, but manganese was rarely detected upon iron breakthrough.

Figure 5.4 summarizes the impact of ferric chloride dose on both the volume treated and run time until breakthrough was observed for the full-scale system under a hydraulic loading rate of 2.2 gpm/ft^2. For this system, a ferric chloride dose of 6 mg/L resulted in a filter run time of about 45 hours with a total volume treated of about 2.4 MG.

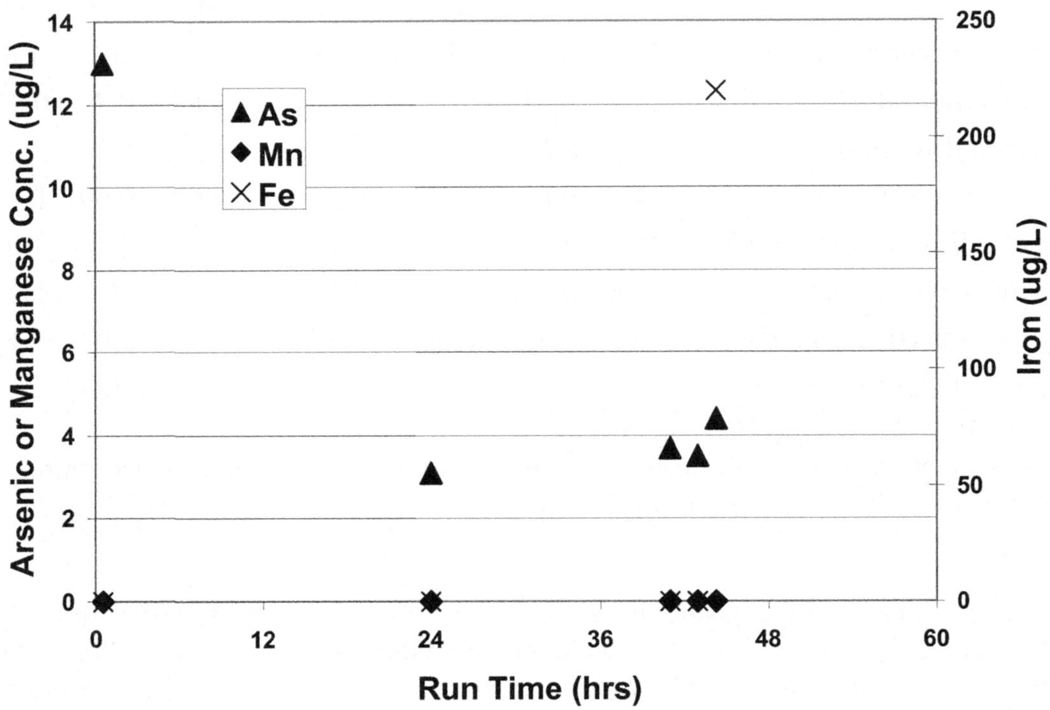

Figure 5.3 Filter Run No. 9 results showing typical observations during breakthrough under 6 mg/L $FeCl_3$ dose (influent As concentration: 14 μg/L)

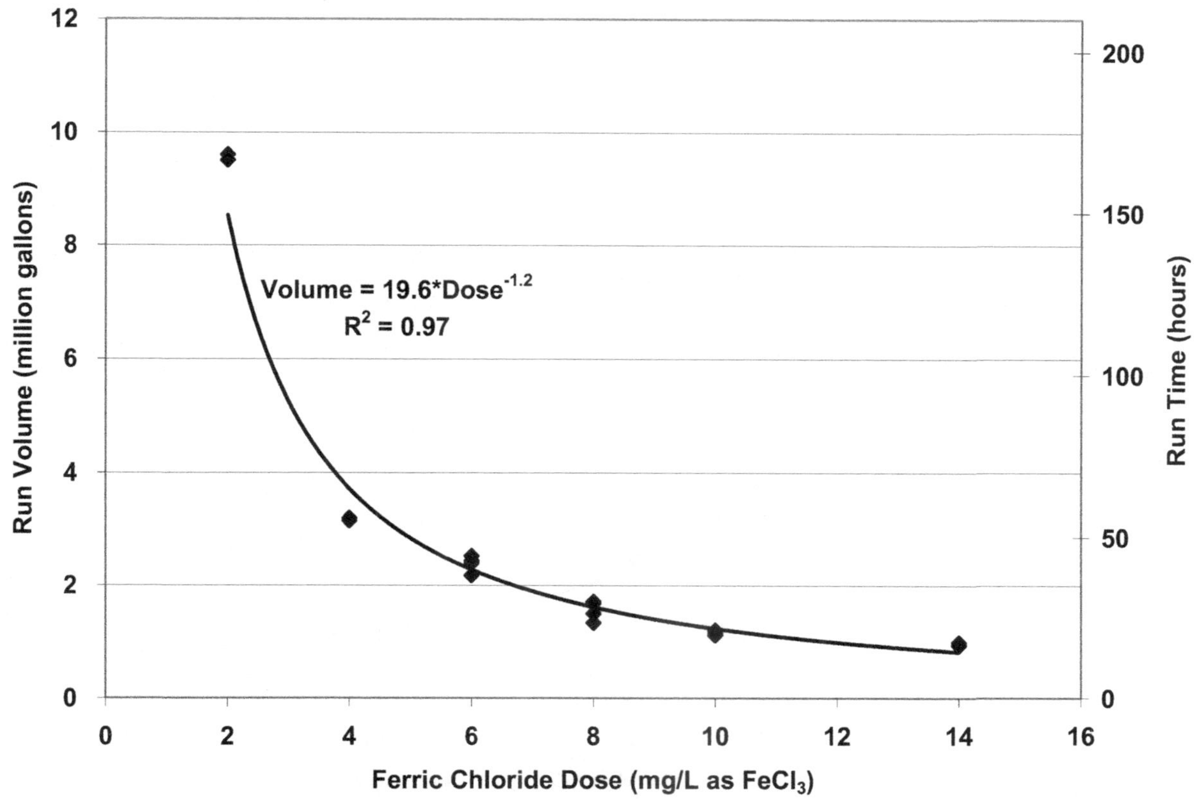

Figure 5.4 Impact of ferric chloride dose on both the volume treated and run time until iron breakthrough was observed for the full-scale system

The overall trend of the data was expected to be linear, but the run time and volume treated for the 2 mg/L dose did not plot on the linear trend established by the greater doses. Only one 2 mg/L dose run was performed through to breakthrough. Another 2 mg/L dose run lasted for 144 hours filtering 6.8 MG while still producing filtered water with non-detectable iron and manganese. Although not completed until breakthrough, this additional testing confirms the non-linear impact of dose on run time and volume treated at the lower doses.

As described in Chapter 3, the concept of UFRV can be useful in translating results for specific treatment conditions to other alternative conditions of filter size and loading rate. The UFRV is the volume treated prior to breakthrough divided by the filter surface area in units of gallons per square foot. The results from the full-scale testing in this project are presented in terms of UFRV in Figure 5.5. Although the UFRV versus ferric dose curve may be system specific, the highly correlated curve shown in Figure 5.5 suggests that once the curve is established for a given system, the filter run time could be accurately predicted when the coagulant dosage needs to be changed in response to fluctuating influent arsenic concentrations or to a different arsenic treatment level required by future regulations. The UFRV observed in the full-scale system correlates reasonably well with those values observed in the pilot-scale testing at similar ferric chloride dosages.

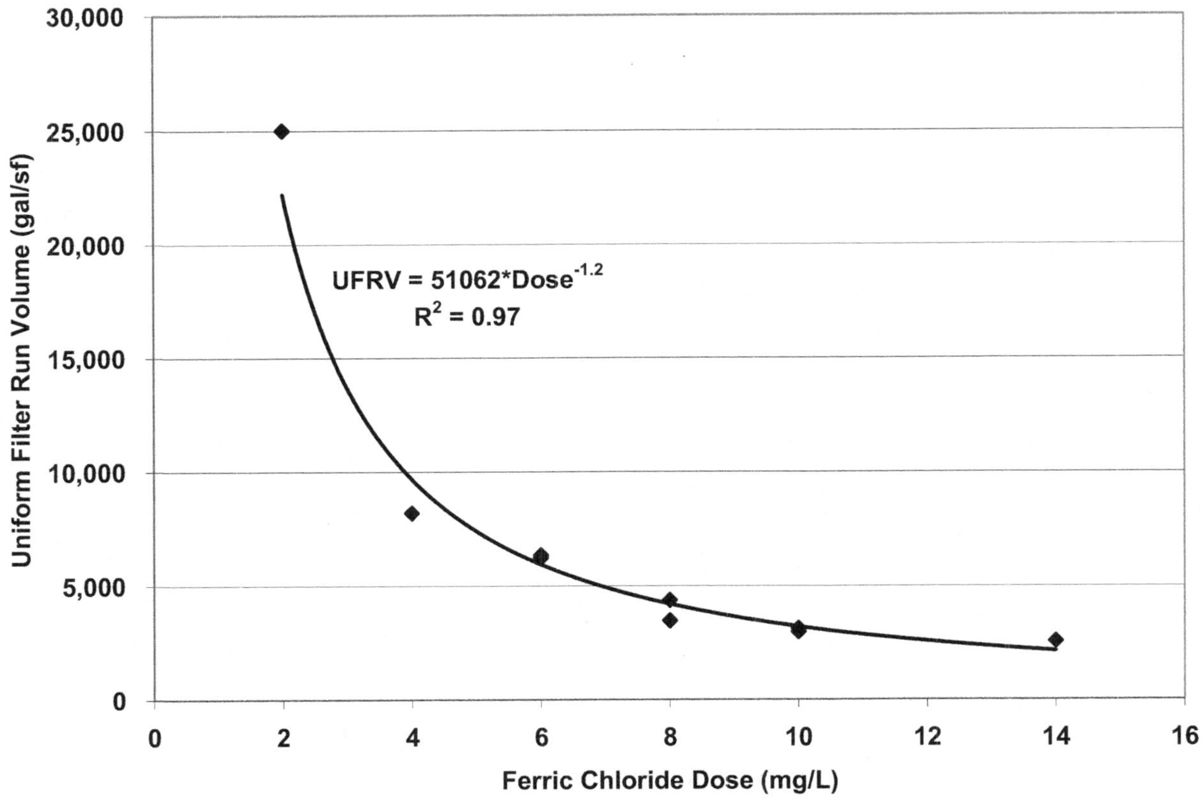

Figure 5.5 Impact of ferric chloride dose on uniform filter run volume

Feedwater recovery is the percentage of water produced relative to the amount of raw water supplied for treatment. The difference between the raw water supplied and the amount of water produced is usually the amount of water used to backwash the filters that is then wasted. A system that uses a backwash supply rate of 12 gpm/ft^2 over a duration of 10 minutes produces 120 gal/ ft^2 of wastewater during backwashing.

Combining this backwashing rate with the UFRVs results in prediction of the feedwater recoveries as a function of ferric chloride dose presented in Figure 5.6. Feedwater recoveries appear to be inversely proportional to ferric chloride dose in accordance with the linear model displayed on the figure. This indicates that feedwater recovery may be reasonably predicted if the required ferric chloride dose is known.

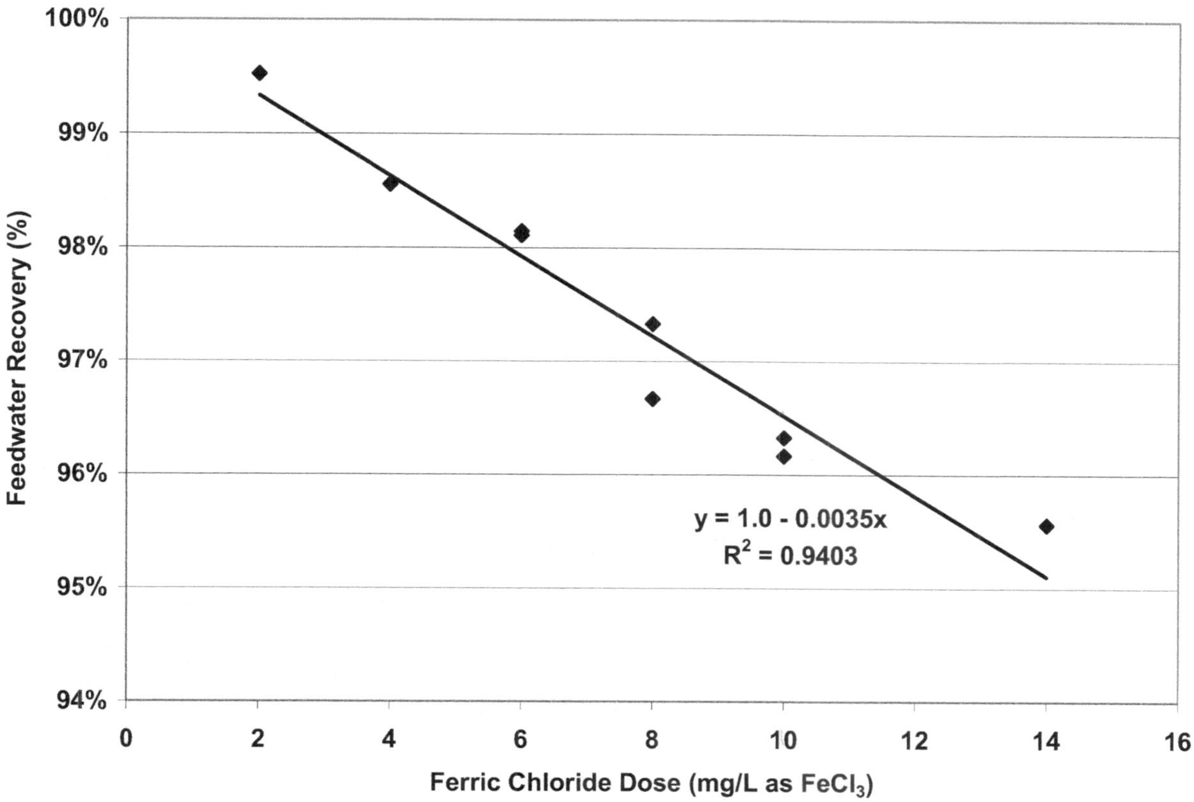

Figure 5.6 Impact of ferric chloride dose on feedwater recovery

Predictive Capability of Process Simulation Methods

Bench- and pilot-scale tests were found to effectively predict filtered water arsenic concentrations when compared with the full-scale results. Figure 5.7 compares results from bench-, pilot-, and full-scale testing. As shown in the figure, the arsenic removal trends between different treatment simulation techniques are very similar. This indicates that utilities can complete either bench- or pilot-scale studies of this process and proceed with process design and budgeting with confidence. However, bench-scale testing provides no predictive capability for filter run length or volume, and cannot be used to optimize the hydraulic loading rate. Pilot-testing is preferred when trying to simulate manganese removal because the media can be conditioned to perform similarly to full-scale systems.

Pilot-testing effectively predicts filter run volumes and times when removing arsenic, iron, and manganese with the process investigated. Figure 5.8 shows the effectiveness with which pilot-scale can predict full-scale run times and volumes for a 2.2 gpm/ft^2 loading rate and ferric chloride dose of 4 mg/L.

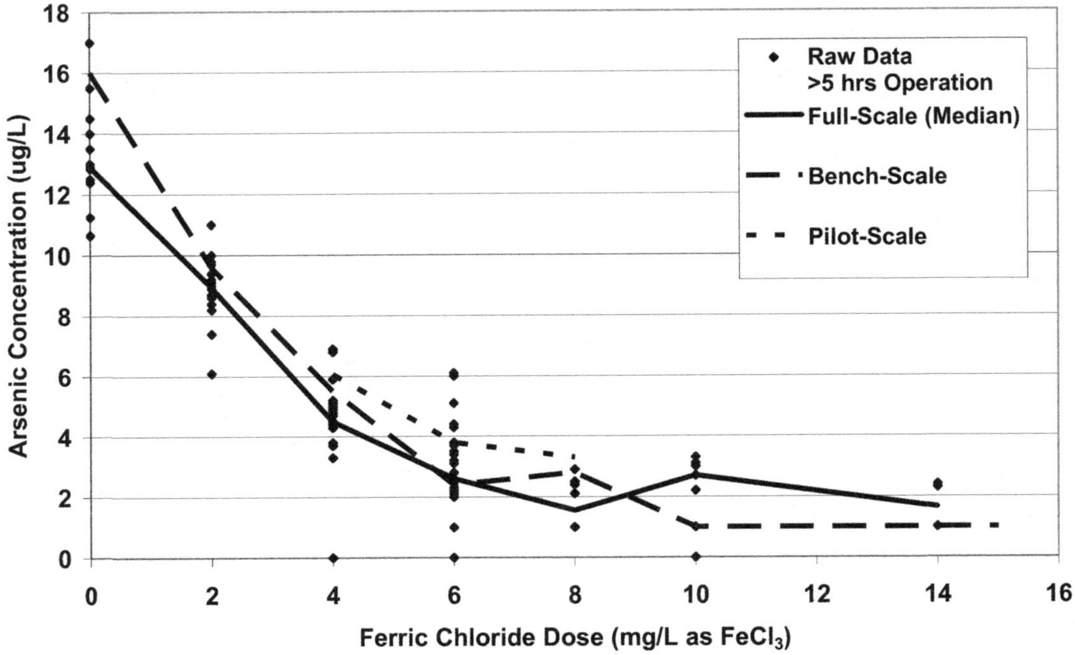

Figure 5.7 Bench- and pilot-scale effectively predicts full-scale filtered water arsenic concentrations

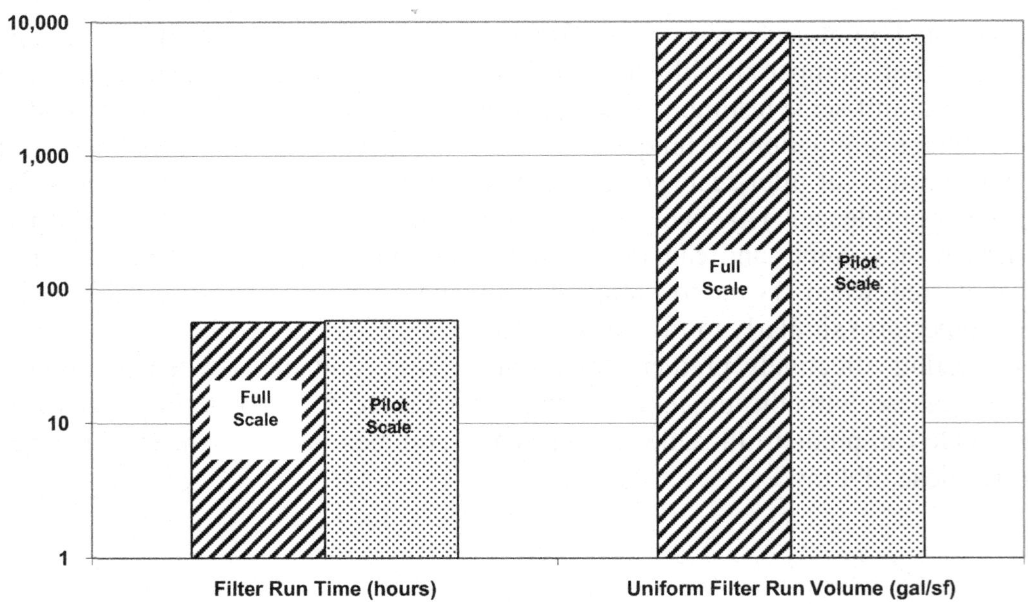

Figure 5.8 Comparison of filter run times and UFRV between pilot- and full-scale results

INFLUENCE OF BACKWASH WATER QUALITY

High initial arsenic concentrations were observed at the beginning of the filter runs where samples were taken within the first four hours of treatment. Filter Run No. 21 shown in Figure 5.9 is a typical example. Arsenic concentrations declined from 12 to 6 μg/L during the first four hours of the filter run. With the exception of the first sample, iron and manganese were not detected in filtered water during the initial four-hour period. Additional investigations were performed to determine the cause of the higher arsenic concentrations in the early part of the filter run, with particular emphasis on whether this would occur during longer-term steady state operations thereby impacting the compliance reliability of the treatment process. Two reasons for the initial high arsenic concentrations were postulated for further investigation:

1. Backwash supply water contains elevated arsenic concentrations and the arsenic-laden backwash water remaining in the filter gradually moves out of the filter once the filter was resume as to forward operation; **or**
2. The coagulation/filtration process does not effectively remove arsenic during the initial part of the filter run due to the lack of filter ripening.

Additional testing was conducted to determine the source for the high initial arsenic concentrations observed at the beginning of Filter Run No. 21 and during other filter runs performed during this study.

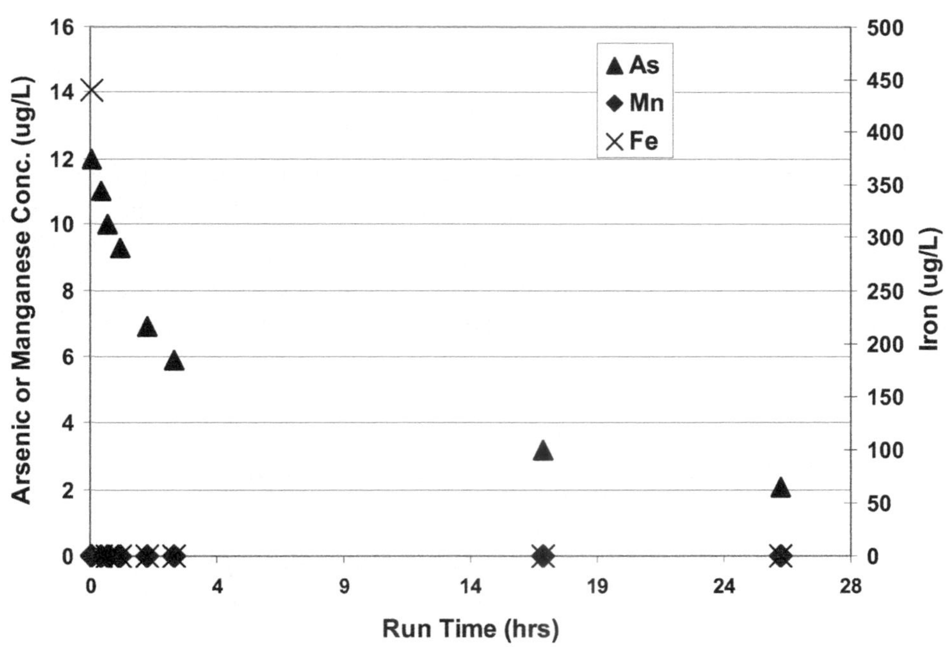

Figure 5.9 Filter Run No. 21: Filtered water quality for 6mg $FeCl_3$/L dose

TESTING PROCEDURE

Additional testing was conducted to evaluate the impact of backwash supply arsenic concentration on arsenic concentrations observed during the initial part of a filter run. Backwash water was supplied from the finished water distribution system. During the overall study, the ferric chloride dose varied considerably for the testing. For instance, doses started at 2 mg/L and were increased by 2 mg/L during each successive run. This resulted in production of filtered water with varying concentrations of arsenic that was then used to supply backwash water. Because the ferric chloride feed system could not be monitored closely on the weekends, it was simply turned off while the system still operated to remove only iron and manganese. Therefore, backwash supply water could contain as much as 16 µg/L of arsenic, the same concentration in the raw water. To investigate the impact of varying arsenic concentrations in backwash supply water, the following two filter runs were performed after backwashing with water that contained different arsenic concentrations (Table 5.5).

The system was operated for 2 days without ferric chloride prior to Filter Run No. 30. Prior to Filter Run No. 30, filters were backwashed with water containing 16 µg/L arsenic. During Run 30, ferric chloride was fed at 6 mg/L for two days. Prior to Filter Run No. 31, filters were backwashed with water containing 2.6 µg/L arsenic. Testing and monitoring conditions included the following.

- Complete 2 runs with a 6 mg/L ferric chloride dose
- Conduct one run with untreated backwash water (Filter Run No. 30) and one run with treated backwash water (Filter Run No. 31).
- Test full-scale system for arsenic, manganese, iron, and chloride
- Sample every 15 minutes for first hour
- Sample every 30 minutes for hours 2-4
- Sample every hour for hours 5-8

RESULTS

Filter Run No. 30, backwashed with water containing 16 µg/L arsenic, exhibited elevated filtered water arsenic concentrations during the first eight hours of the filter run (see Figure 5.10). These results are similar to Filter Run No. 21 discussed previously. Filter Run No. 31, which was backwashed with water containing low arsenic concentrations, did <u>not</u> exhibit elevated finished water arsenic concentrations during the initial part of the filter run (see Figure 5.11).

Table 5.5
Backwash supply water arsenic concentrations

Filter Run	Backwash supply water arsenic concentration (µg/L)
30	16
31	2.6

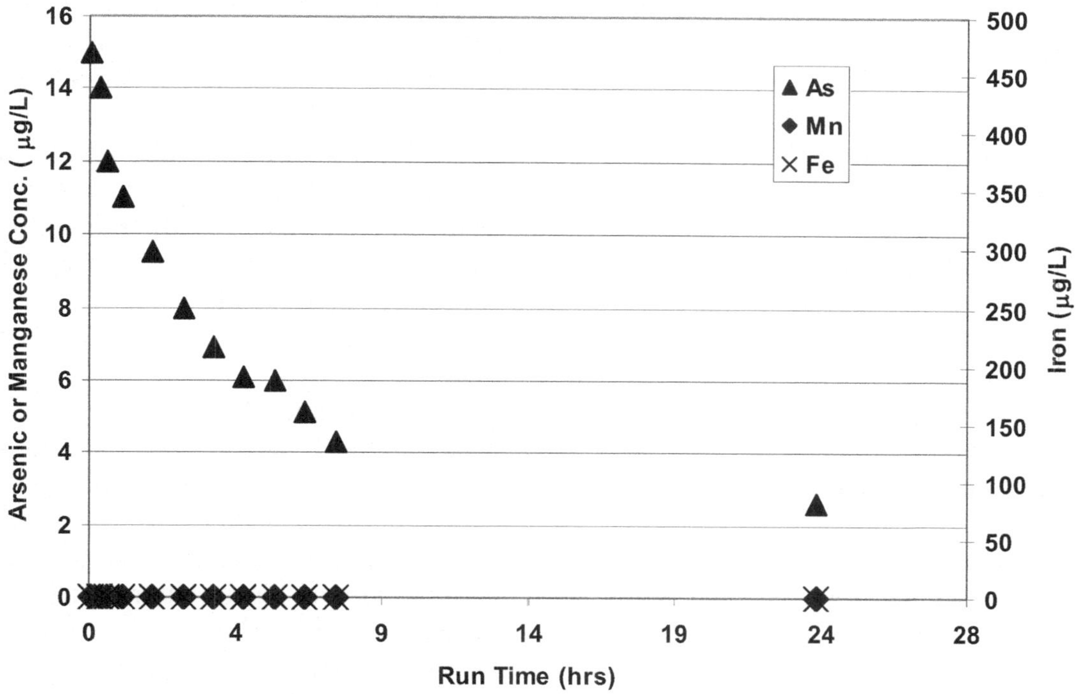

Figure 5.10 Filter Run No. 30: Filtered water arsenic concentrations after backwashing with water containing 16 µg/L arsenic

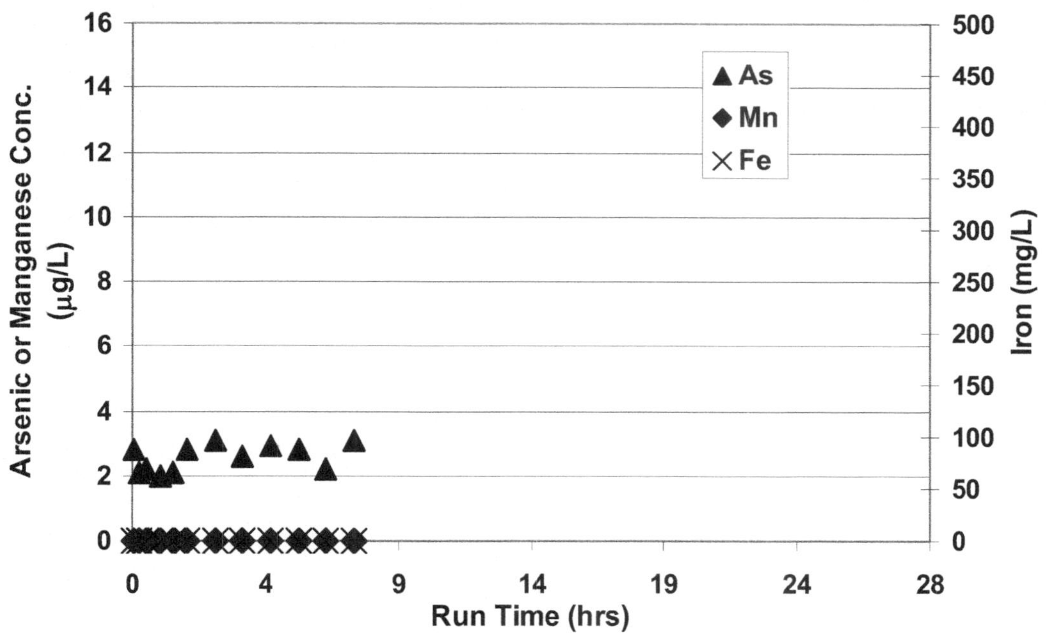

Figure 5.11 Filter Run No. 31: Filtered water arsenic concentrations after backwashing with water containing 2.6 µg/L arsenic

Comparison of results from Filter Runs No.30 and No.31 convinced the investigators that the high concentrations during the initial portion of the filter run resulted from high concentrations of arsenic in the backwash supply water. Because ferric chloride doses were varied during this study, the backwash supply often contained higher concentrations of arsenic than the filtered water. This results in artificial increases in filtered water arsenic during the initial part of the filter runs. During long-term steady state operation, the concentration of arsenic in the backwash supply will be the same as the filtered water, resulting in uniform arsenic concentrations in filtered water during the filter run as observed in Filter Run No. 31 (Figure 5.11). Therefore, utilities using this treatment process to consistently achieve a target arsenic concentration will not observe the same initial high concentrations of arsenic in the beginning of a filter run.

Filter Run No. 30 (Figure 5.10) exhibited elevated arsenic concentrations for an extended period of eight hours or more. This seemed to be much longer than needed to simply hydraulically flush out the backwash supply water, even when considering dispersion within the filter. To investigate this, chloride was also monitored during the filter runs to serve as a tracer. The ambient water contains chloride, but the ferric chloride provides an additional 20 percent. Prior to Filter Run No. 30, the system was operated without ferric chloride addition so the backwash supply was expected to contain only the ambient chloride concentration. After backwashing, the forward flow of the filter would have the additional 20 percent chloride added by injecting ferric chloride. Therefore, some understanding of the hydraulics of the filter can be obtained by observing the increase in filtered water chloride concentrations as the backwash supply water is displaced by ferric chloride-dosed water moving through the filter.

Figure 5.12 presents the normalized concentration of chloride during the filter run. The normalized concentration of chloride begins at ambient concentration, but increases to a steady state of about 20 percent above ambient in 2.5 hours. Therefore, it is estimated that it takes about 2.5 hours for all of the backwash supply water to exit the filter.

However, as shown in Figure 5.12, arsenic concentrations continue to decline even after the chloride reaches steady state and the backwash supply water is expected to be flushed from the filter. Because arsenic is retained in the filter longer than the backwash supply water, it appears that the declining arsenic concentrations cannot be explained simply by hydraulic flushing of the backwash water out of the filter at the start of forward flow.

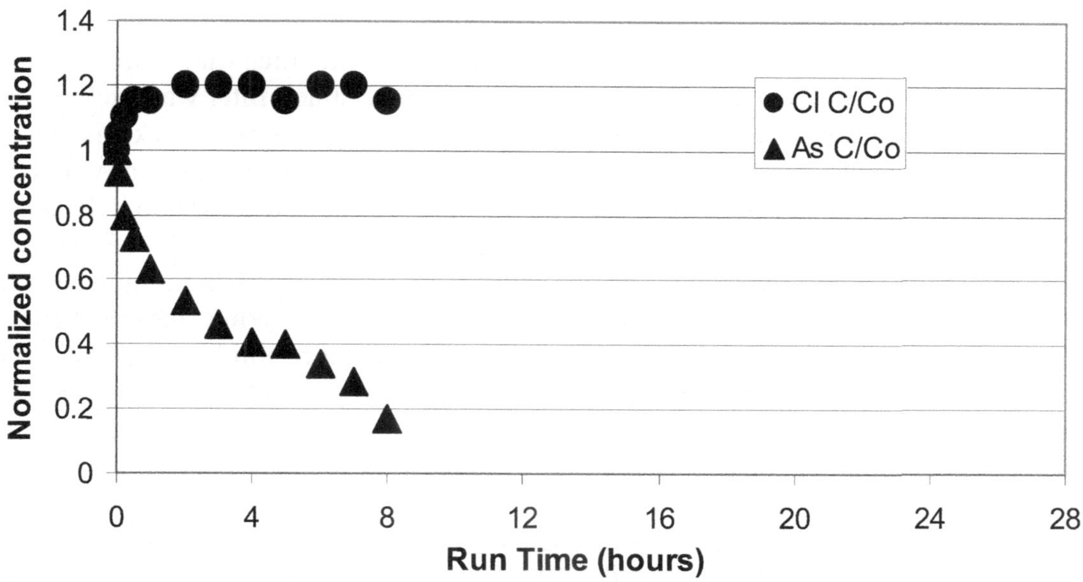

Figure 5.12 Comparison of chloride and arsenic behavior during Filter Run No. 30

One possible explanation for this phenomenon is that arsenic may be retained in the filter longer than the backwash water due to adsorption. Arsenic may adsorb to the surface of filter media (which is comprised of manganese oxides and possibly some residual ferric hydroxide solids) during backwash to achieve a solid phase concentration commensurate with the liquid phase concentration of 16 μg/L. Then, when forward flow is initiated and water is being produced with lower arsenic concentrations, arsenic may desorb from the solid to the liquid phase, thereby increasing filtered water arsenic concentrations. Because this may not be an instantaneous process, it may explain the rather lengthy decline in arsenic concentrations during the start of the filter run. If sorption/desorption is the mechanistic explanation for these observations, then high initial arsenic concentrations during the filter run could occur anytime after the ferric chloride feed has been interrupted. Examination of arsenic sorption/desorption in filters for iron and manganese removal may be a good topic for further research to refine this treatment process and examine its reliability for compliance with the arsenic MCL.

SLUDGE HANDLING

Waste disposal is an important consideration in arsenic treatment. The primary wastes generated by oxidation, coagulation, and filtration are related to filtration backwash waste. Backwash waste (BWW) is a dilute liquid waste stream that contains solids, including iron oxides, manganese oxide solids, and natural colloids or particulates in the well water, and arsenic that was removed from the raw water and adsorbed onto the solids (primarily on iron oxides). This waste stream may either be disposed directly to a sewer system or further processed to create solid waste. These wastes have the potential for being classified as hazardous and can pose disposal problems.

Backwash Waste Treatment

The pressure filters were backwashed following each filter run. Finished water in the distribution system was used as the backwash water supply. Each filter was backwashed for four minutes at 1,000 gpm following a 30 second surface wash at 320 gpm. Backwash waste was de-chlorinated by injecting sodium thiosulfite. Backwash waste was captured in two 20,000-gallon temporary backwash water holding tanks and was allowed to settle for about 10 hours prior to decanting. Approximately 30,000 gallons was decanted from the tanks after settling with 10,000 gallons of settled sludge retained. The supernatant was discharged to the adjacent stormwater drainage channel under the existing NPDES discharge permit. While more solids were being added to the holding tank after each run, the settling of the entire solid blanket had become more efficient and relatively rapid. All solids settled at the bottom of the tank at the end of each settling, with the solid-liquid separation line always below the bottom 10,000 gallon mark. Approximately 10,000 gallons of settled liquid sludge was generated during the entire course of the project.

Quarterly testing was required for NPDES compliance with the existing permit, and the decant water quality consistently met the NPDES discharge limitations. The modified system discharged better water quality than the existing system, due to settling of the backwash waste water. Decant water quality results are presented in Table 5.6. Arsenic concentrations in the discharge water were observed to be 14 μg/L (8/26/02) and 8.8 μg/L (9/03/02). Due to intermittent operation of the ferric chloride feed system, the finished water (backwash supply) had arsenic concentrations in the range of 8 to 17 μg/L. Upon consistent operation of the ferric chloride feed system in the future, the backwash supply would always have arsenic concentrations less than 10 μg/L and a consequent reduction in discharge water arsenic concentration may be observed.

Table 5.6
Decant water quality sampling results (July 10, 2002)

Constituents	Decant		Discharge Limitations	
	From BWW Settling	Units	Monthly Average	Daily Maximum
pH	7.8	std Units		6.0-9.0
Iron	<100	μg/L		
Manganese	<20	μg/L		
Temperature	72	°F		100°F
Oil and grease	<2.0	mg/L	10	15
Settleable solids	<0.1	ml/L	0.1	0.3
Total suspended solids	3	mg/L	50	150
Turbidity	0.3	NTU	50	150
Sulfides	<1.0	mg/L		1.0
$BOD_5$20°C	<5.0	mg/L	20	30

Sludge Production and Disposal

Approximately 10,000 gallons of settled sludge was produced during this project where 90.7 million gallons of water was treated with an average ferric chloride dose of 4.4 mg/L. Summary information regarding the mass of material removed during the study and the resultant waste characteristics is provided in Table 5.7. Due to the ferric chloride addition, iron was by far the greatest mass removed at 1,550 kg over the course of the study. Thirty killograms of manganese and 2.6 kg of arsenic were also removed. The expected sludge concentration (after settling and decanting) over the duration of the study was predicted to be 70 mg/L of arsenic. Once the sludge was dried, iron was predicted to be the dominant element, but arsenic concentration was predicted to be 1,667 mg/Kg on a dry weight basis exceeding the 500 mg/kg Total Threshold Limit Concentration (TTLC, California criterion). The overall ratio of arsenic removed per iron added was determined to be in the range of 6 μg of arsenic per mg of iron. Using this ratio and assuming ferric hydroxide $Fe(OH)_3$ is the predominant solid arsenic content of the dry sludge, the overall arsenic content would be expected to be in the range of 3,000 mg As/Kg, assuming Fe comprises 50 percent of the weight of ferric hydroxides. Considering many runs in this study were operated with ferric chloride dosages higher than 6 mg/L, which could, in turn, reduce the arsenic sorption density of ferric hydroxides, the analyzed arsenic concentration in the dry sludge, 1,667 mg As/Kg, was in the same ballpark as that estimated by the arsenic sorption density, which is 3,000 mg As/Kg. Utilities that are considering using coagulation/filtration technology for arsenic compliance should carefully estimate the arsenic concentration in the backwash wastewater and settle sludge so that these waste streams will not exceed the threshold of being hazardous. This will be particularly important for those who will incorporate equalization tanks for backwash wastewater.

Table 5.7
Mass removed and predicted sludge characteristics

	Fe	Mn	As
Native mass removed (kg)	25.1	30.6	2.6
Ferric chloride mass removed (kg)	1,524.8	N/A	N/A
Total mass removed (kg)	1,549.8	30.6	2.6
Expected sludge concentration (mg/L)	40,946	807	70
Expected dry sludge composition (%)	97.9	1.9	0.2
Expected dry sludge composition (mg/kg)	979,031	19,302	1,667

Notes: Sludge volume (gal) 10,000

As discussed previously in this chapter, arsenic removal increases with increasing ferric chloride dose. Therefore, a greater mass of arsenic is removed during a filter run with a greater dose. However, the feedwater recovery decreases with increasing ferric chloride dose (more backwash waste is generated due to shorter filter runs). The combination of greater arsenic removal and greater backwash water use with increasing ferric chloride dose tend to make the expected backwash water arsenic concentration relatively independent of ferric chloride dose. This assumes that filter runs are operated until breakthrough and that arsenic removal follows the ratio described earlier (6 μg As:mg Fe).

For the conditions investigated in this study, computed average arsenic concentrations in the backwash waste (after each run) were between 1.5 to 2.0 mg/L and these numbers were relatively independent of ferric chloride as illustrated in Figure 5.13. If direct disposal of backwash waste (such as sewer discharge) is desirable, maintaining a backwash wastewater arsenic concentration below 5 mg/L is desirable due to hazardous waste limits discussed in the following section of this chapter.

The overall recommended process for handling and disposing of backwash wastes generated from oxidation, coagulation, and filtration is equalization, settling, recycle of supernatant to the treatment process, and discharge of the settled sludge to the sanitary sewer system. However, this process must be engineered for operation within regulatory criteria, and the process may not be feasible for all systems.

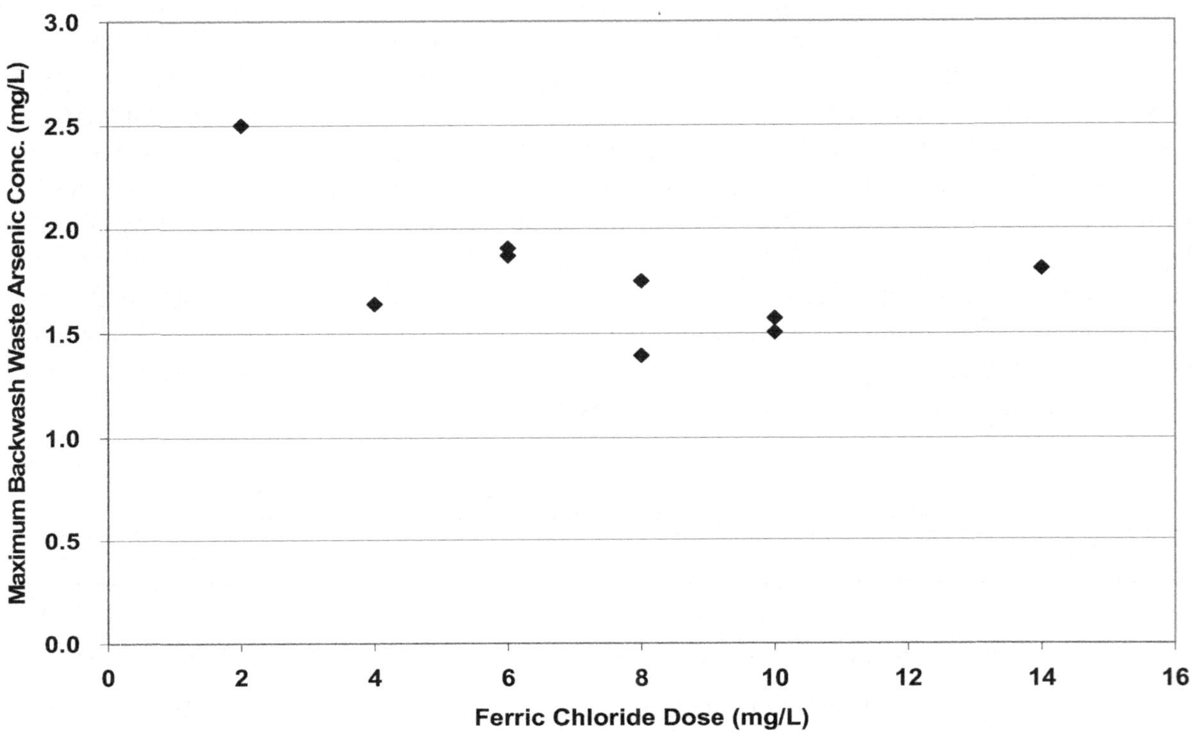

Figure 5.13 Impact of ferric chloride dose on expected maximum backwash waste arsenic concentration

Disposal Regulations

Waste disposal in California involves both compliance with the provisions of the federal Resource Conservation and Recovery Act (RCRA) and compliance with state regulations. The waste must be characterized as solid or liquid. To be classified as a solid by the Los Angeles Sanitation District, the waste must be greater than 50 percent solids and have no free-standing water measured by the paint filter test (PFT).

Liquids waste having an arsenic concentration of more than 5 mg/L will be classified as hazardous waste. Liquid wastes can be discharged directly (must comply with the Clean Water Act [CWA] and Non-hazardous NPDES) or indirectly through the sanitary sewer system (must comply with TBLLs).

In California, a solid waste is considered hazardous if its arsenic concentration [Total Threshold Limit Concentration (TTLC)] is greater than 500 mg/kg. If not hazardous based on the TTLC, solid waste is subject to the Toxicity Characteristic Leaching Procedure (TCLP, federal requirement) and, if in California, a Waste Extraction Test (WET). If the concentration of arsenic in the TCLP leachate is greater than 5 mg/L the material is characterized as hazardous waste and would require hazardous waste disposal procedures. If the waste passes the TCLP but not the WET, it is classified as a non-RCRA hazardous waste in California. Non-RCRA wastes are treated as hazardous in California per state regulations but are not hazardous in other states that do not have requirements more strict than the federal RCRA requirement.

The following list provides a summary of the available waste disposal options, governing regulation, and associated testing. Figure 5.14 shows the flow diagram for the treatment sludge, and illustrates a decision process relative to the classification of arsenic waste and associated disposal options.

Liquid (percent solids, PFT)

- Direct Discharge (CWA, NPDES)
- Indirect Discharge through Sanitary Sewer (TBLLs)

Solid (percent solids, PFT)

- Land Application (LACSL)
- Conventional Landfill (PFT, TCLP, WET)
- Hazardous Landfill (PFT)

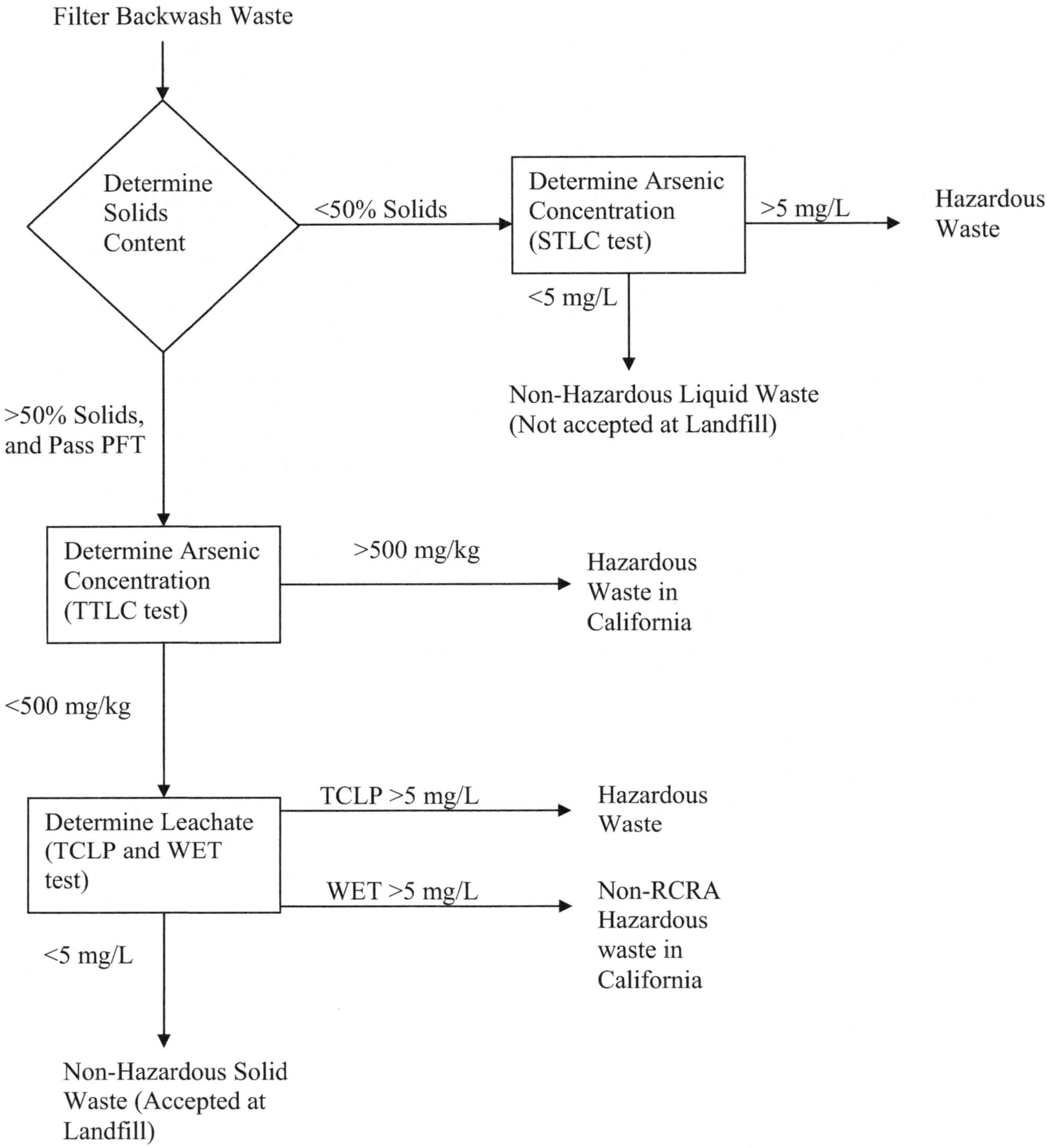

Figure 5.14 Regulatory classification of arsenic filtration treatment wastes

CHAPTER 6
COST ANALYSIS

Costs for building an oxidation, coagulation, and filtration process are site-specific based on site conditions and owner preferences for equipment and housing. However, based on the information developed in this project, costs were estimated for two plant sizes and results were used to develop a general construction cost curve. Detailed cost estimates are provided in Appendix E to show the equipment required in a project of this type. The general cost estimates can be used as a basis to begin a site-specific cost estimate for a specific application. Any use of the cost information presented should be considered within the context of the process design and criteria upon which the estimates are based.

Costs presented in this chapter are based on a hydraulic loading rate of 5 gpm/ft^2. Changes to the hydraulic loading rate may have a dramatic impact on the size of filters and the ultimate cost of the facility. The residuals handling process is assumed to be a batch process including equalization, settling, decant, and sludge discharge to the sanitary sewer. From an operation and maintenance (O&M) cost perspective, the ferric chloride feed rate is assumed to be 6 mg/L. The resulting cost curves are presented in Figure 6.1, and detailed estimates along with O&M cost information is presented in Appendix E.

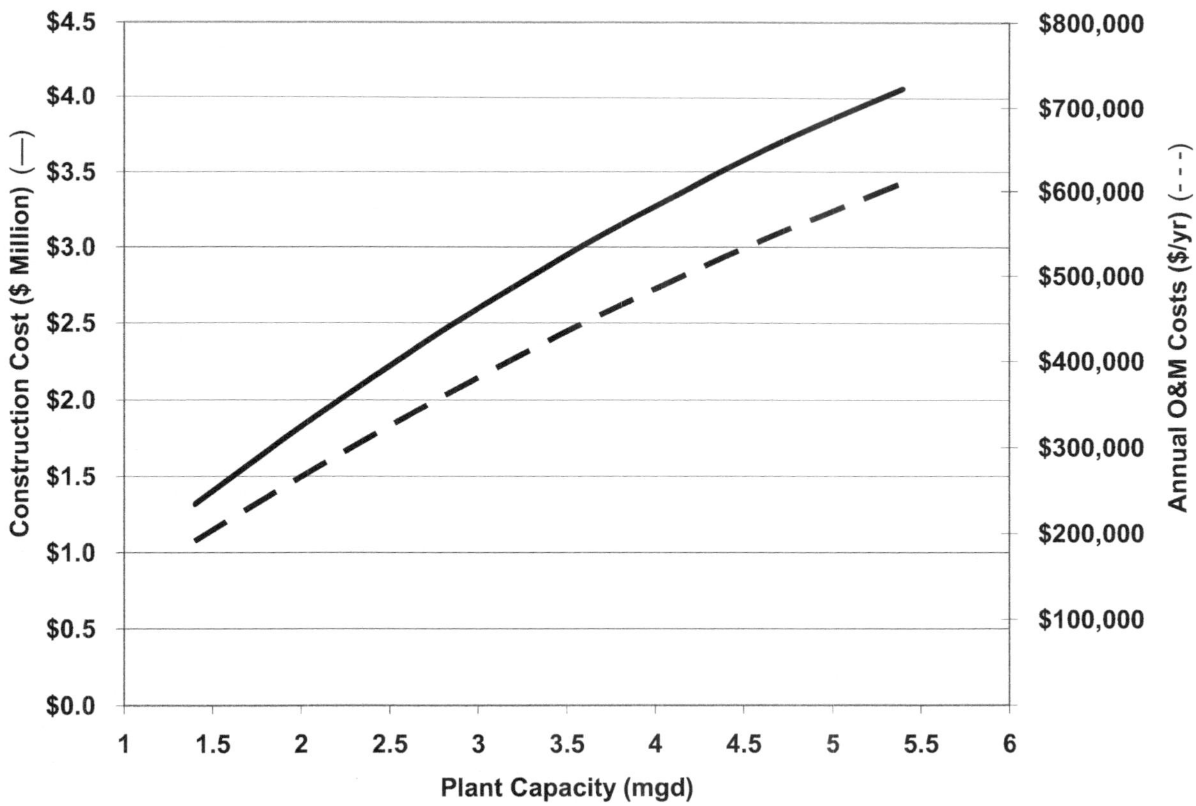

Figure 6.1 Impact of plant capacity on expected costs

CHAPTER 7
CONCLUSIONS

Many utilities that need to treat arsenic to be in compliance with the new arsenic MCL also need to address other contaminants that are either on the primary or secondary MCL list. In fact, a significant number of utilities already have traditional treatment facilities in place for iron and manganese removal. Theoretically, it is possible to retrofit these existing facilities with ferric addition to accomplish simultaneous arsenic and manganese removal. While the treatability of this concept has been proven in the conventional coagulation/filtration setting, questions and concerns remained for the incorporation of ferric addition to a manganese removal plant. This study provides utilities with detailed implementation information and treatment simulation methods that can be used to effectively determine coagulant dose, filter run time, and waste generation volumes for site-specific applications. Major findings from this research project are summarized as follows.

1. Addition of ferric chloride to the existing oxidation/filtration process results in adequate arsenic removal without sacrificing manganese removal performance.
2. The implementation of ferric chloride addition is a fairly simple and straightforward process with minimal additional equipment and instrumentation requirements.
3. For the SCWC Centralia site that typically had 16 μg/L influent arsenic, filtered water arsenic concentrations as low as non-detect (<2 μg/L) could be achieved, depending on the ferric chloride dose.
4. A ferric chloride dose of between 2 and 4 mg/L was sufficient to achieve a target filtered water arsenic concentration of 8 μg/L. A ferric chloride dose of 6 mg/L was sufficient to achieve a target filtered water arsenic concentration of 5 μg/L.
5. The addition of ferric chloride shortens filter runs due to the additional solids added, with higher ferric chloride doses resulting in shorter filter runs. At the Centralia site, filter runs were always terminated due to the breakthrough of iron (as opposed to breakthrough of other contaminants or excess headloss). A ferric chloride dose of 4 mg/L was sufficient to reach the target filtered water arsenic concentration of 8 μg/L and resulted in a filter run of 55 hours with a Uniform Filter Run Volume (UFRV) of 8,200 gallons/ft^2.
6. The backwash waste stream did not contain sufficient arsenic concentration to be classified as hazardous waste. However, once concentrated through several batch settling and decant cycles, the retained settled sludge exceeded the 5 mg/L arsenic criterion for liquid residuals. When the liquid waste was dried, it passed the Toxicity Characteristic Leaching Procedure (TCLP) but did not pass California's Waste Extraction Test (WET). Therefore, the solid waste in California would be classified as a non-Resource Conservation and Recovery Act (RCRA) hazardous waste. The solid waste would not have been classified as hazardous outside of California. Due to the residuals characteristics and the lack of space on the Centralia site, the waste can be most readily handled through equalization, settling, decant of the supernatant to the stormwater channel, and discharge of the settled sludge to the sanitary sewer. Costs of the sewer connection were not specifically evaluated but typically include both a connection and use charge with a potential surcharge for strength.

7. Bench-scale testing was found to effectively predict the full-scale testing results for removal of arsenic by ferric chloride and filtration. Bench-scale testing is not expected to effectively simulate manganese removal because its rate of oxidation is impacted by manganese coatings on the filtration media. Bench-scale results do not address filter run lengths or uniform filter run volumes.
8. Pilot-testing effectively simulated the full-scale system. Beyond determining the impact of ferric chloride addition on filtered water arsenic concentrations, pilot testing can effectively predict manganese removal (provided the filter media is seasoned or greensand is used), filter run times, and uniform filter run volumes.
9. Filter loading rates were examined during pilot testing. No correlation was found between filter loading rate and filtered water arsenic concentration for loading rates between 2.2 and 8 gpm/ft^2. The UFRV was slightly lower at higher loading rates. Increasing the loading rate from 2.2 to 8 gpm/ft^2 decreased the UFRV by 10 percent. Because essentially the same volume is produced in the filter run, the filter run length is inversely proportional to the loading rate.
10. Use of backwash water with greater arsenic concentrations than the filtered water was found to artificially increase the arsenic concentrations during the early portions of the filter run. Based on the observation of gradually declining filtered water arsenic concentration at the beginning of the filter run, the filtration media appeared to have a limited capacity for adsorption/desorption of arsenic. Therefore, after exposed to water with higher concentrations of arsenic, the filter media tends to supply dissolved arsenic to the process stream through desorption over the initial eight hours of the run. This should not be a problem for filters that are operated with a uniform ferric chloride dose and are backwashed with filtered water. When backwash water with equivalent or lower concentrations to the filter effluent was used, filtered water arsenic concentrations remained uniform during the filter run.

APPENDIX A
BENCH TEST PLAN

WATER SAMPLING AND BENCH TEST PLAN

Southern California Water System
American Water Works Research Foundation
Simultaneous Arsenic and Manganese Removal

Prepared by: Bryan Black, PE — August 20, 2001
Reviewed by: Yu-Jung Chang, PhD

OBJECTIVES

1) Evaluate impact of ferric chloride dose on arsenic removal
2) Evaluate impact of ferric chloride with pH fixed at ambient
3) Evaluate impact of pH on arsenic removal with sub-optimal ferric chloride dose

WATER SAMPLES (SCWC)

1) Samples for Bench Test (Collect and ship on Monday 8/20)
   a) Collect three, five-gallon samples
      i) Combined well water
      ii) No chlorine
      iii) Attach tubing to sample port
      iv) Adjust sample flow rate to about 1 gallon per minute
      v) Flush sample line for 5 minutes to waste
      vi) Prevent entrainment of air during sampling
      vii) Fill container with discharge tube submerged in water sample
      viii) Fill container completely to eliminate head space (air)
   b) Ship overnight to HDR lab:
      HDR Engineering
      12839 NE 14th Place
      Bellevue WA 98005
      Phone: 425/450-6376
      Attn: Sid Hendrickson
2) Raw water for laboratory analysis (Prepared by SCWC)
   a) Collect whole water sample (raw water direct into sample container)
      i) Submit whole water sample for: arsenic, iron, manganese, TOC, TDS, Sulfate, Phosphate
   b) Collect filtered sample (filter raw water through 0.45 μm filer into sample container)
      i) Submit filtered sample for: arsenic, iron, manganese, silica
   c) Take field measurements of raw water
      i) pH, temperature, turbidity
   d) Arsenic species
      i) Follow instructions in equipment kit provided (Method of Edwards, 1998)
      ii) Collect filtered sample (through 0.1 μm filer into clean container)
      iii) Filter water through ion exchange media
      iv) Submit filtrate for arsenic determination

GENERAL BENCH TEST PROCEDURE

1) Measure raw water pH
2) Collect samples for jar test in 6-place stirrer
3) Turn mixer to 100 rpm
4) Dose chemicals per Table 1 in order listed (Hypochlorite dosed at least 30 seconds prior to ferric chloride)
5) Rapid mix (100 rpm for 30 seconds following complete chemical dosing)
6) Slow mix (20 rpm for 8 minutes)
7) Draw sample from fixed port
8) Vacuum filter through 0.45 μm membrane
9) Send sample to lab for arsenic, iron, manganese
10) Analyze for pH, turbidity

Table A.1
Bench Test Conditions

Sample	Sodium Hypochlorite Dose (mg/L as Cl_2)	Ferric Chloride Dose (mg/L as $FeCl_3$)	pH
Run A – Ferric chloride dose varied, pH not controlled			
1	3	2	uncontrolled
2	3	4	uncontrolled
3	3	6	uncontrolled
4	3	8	uncontrolled
5	3	10	uncontrolled
6	3	15	uncontrolled
Run B – Ferric chloride dose varied, pH controlled to raw ambient			
7	3	2	Fixed at raw ambient
8	3	4	Fixed at raw ambient
9	3	6	Fixed at raw ambient
10	3	8	Fixed at raw ambient
11	3	10	Fixed at raw ambient
12	3	15	Fixed at raw ambient
Run C – Ferric chloride dose fixed at 4 mg $FeCl_3$/L, pH varied			
13	3	4	8.0
14	3	4	7.5
15	3	4	7.0
16	3	4	6.5
17	3	4	6.0
18	3	0	Uncontrolled

Titration

The experiments with pH control will require a titration prior to the jar test to determine the quantities of acid or base to be added. Water remaining from Run No.1 can be used for the titration. Sodium hydroxide will be used to increase pH. Sulfuric acid will be used to decrease pH.

Ferric Chloride Stock Solution

Ferric chloride is 39% by weight with specific gravity of 1.4. It contains 546 mg FeCl3/mL solution.

$$\frac{1.4g\_soln}{mL\_soln} * \frac{0.39g\_FeCl_3}{g\_soln} * \frac{1{,}000mg}{g} = 546 \text{ mg } FeCl_3\text{/mL soln.}$$

Add 7.3 mL neat ferric chloride to 1-L volumetric flask. Fill to mark with DI water. Solution strength is approximately 4,000 mg FeCl3/L .

$$\frac{7.3mL\_soln}{1L} * \frac{546mg\_FeCl_3}{mL\_soln} = 3985.8 \text{ mg } FeCl_3\text{/L}$$

One mL of stock solution added to 2-L sample is equivalent to 2 mg FeCl3/L dose.

$$\frac{3985.8mg\_FeCl_3}{L} * \frac{1mL}{2-L} * \frac{L}{1{,}000mL} = 1.99 \text{ or } 2 \text{ mg } FeCl_3\text{/L dose}$$

Lab Information

See Yu-Jung Chang for sample containers (1-pint plastic). Samples do not need to be preserved, do not need ice, and do not need to be shipped overnight. Mark samples as "not preserved". Analysis of iron and manganese by 200.7-ICP. Analysis of Arsenic by SM3113B GFAA.

Clinical Lab
21881 Barton Rd.
Grand Terrace, CA 92313
(909) 825-7693
Fax (909) 825-7696
Dr. Mehdi Siami, Director
or
Bob Glaubig

Detection limits are Iron 100 µg/L, Manganese 20 µg/L, and Arsenic 2 µg/L.

Calculation of Slow Mix Conditions

In the full-scale plant, water will first pass through the static mixer and then mix slowly in piping to the pressure filters as well as in the empty volume above the filter media. The jar test will simulate this slow mixing. The maximum full-scale water production rate is 1,150 gallons per minute. After static mixing, water flows through about 60' of 12" pipe and 30' of 8" pipe. Filter media occupies the lower one-half of the pressure vessel and the empty upper portion of the pressure vessel provides some additional contact time for floc formation. Velocities in the 12" and 8" pipes are 3.3 and 7.3 feet per second. Contact times in the 12" and 8" pipe are 18.2 and 4.1s. The upper one-half of the pressure filters has a volume of 9,020 gallons with a contact time of 7.8 minutes. Therefore, the slow mix step should proceed at 20 rpm for 8 minutes.

APPENDIX B
PILOT TESTING RESULTS

Table B.1
Summary of pilot scale test results

Run Number	Start Date	Filter Material	Dose (mg/L FeCl3)	Loading Rate (gpm/ft2)	Break through Reached? (y/n)	Chart Location
7	8/15/2002	dual media	8	5	y	B-3
8	8/19/2002	dual media	4	2.2	y	B-3
9	8/22/2002	dual media	4	8	y	B-4
10	8/26/2002	dual media	4	5	y	B-4
11	8/28/2002	dual media	6	5	y	B-5
12	9/3/2002	dual media	6	3	y	B-5

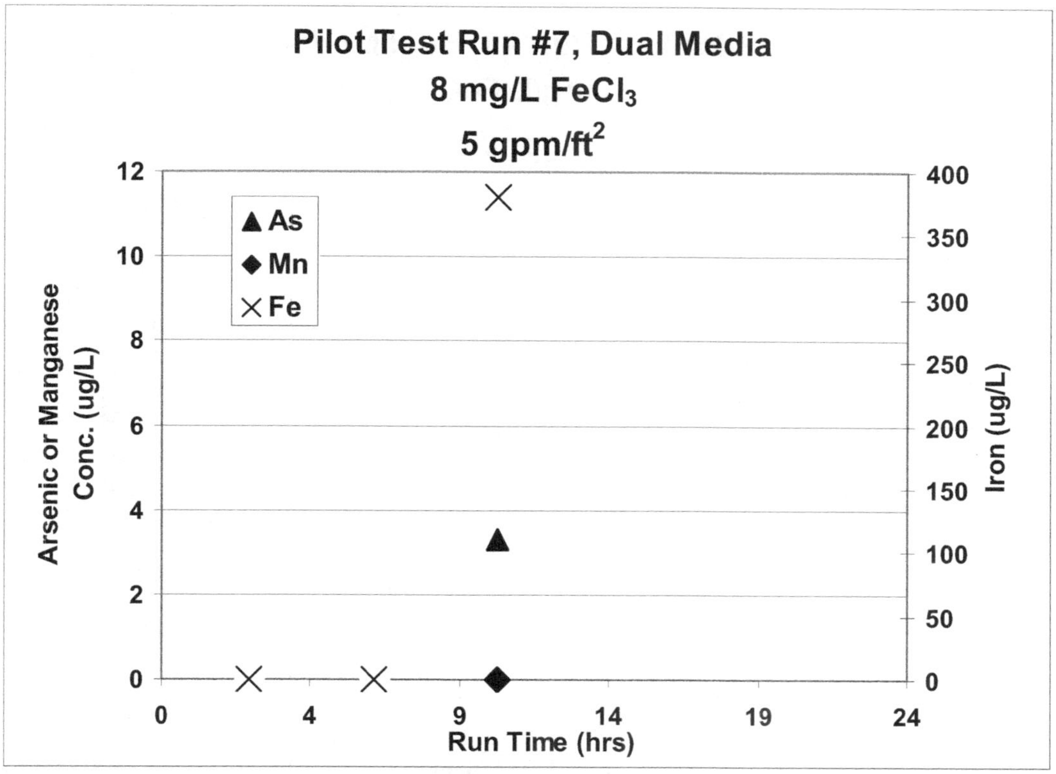

Figure B.1 Pilot test run #7, dual media, 8 mg/L $FeCl_3$, 5 gpm/ft^2

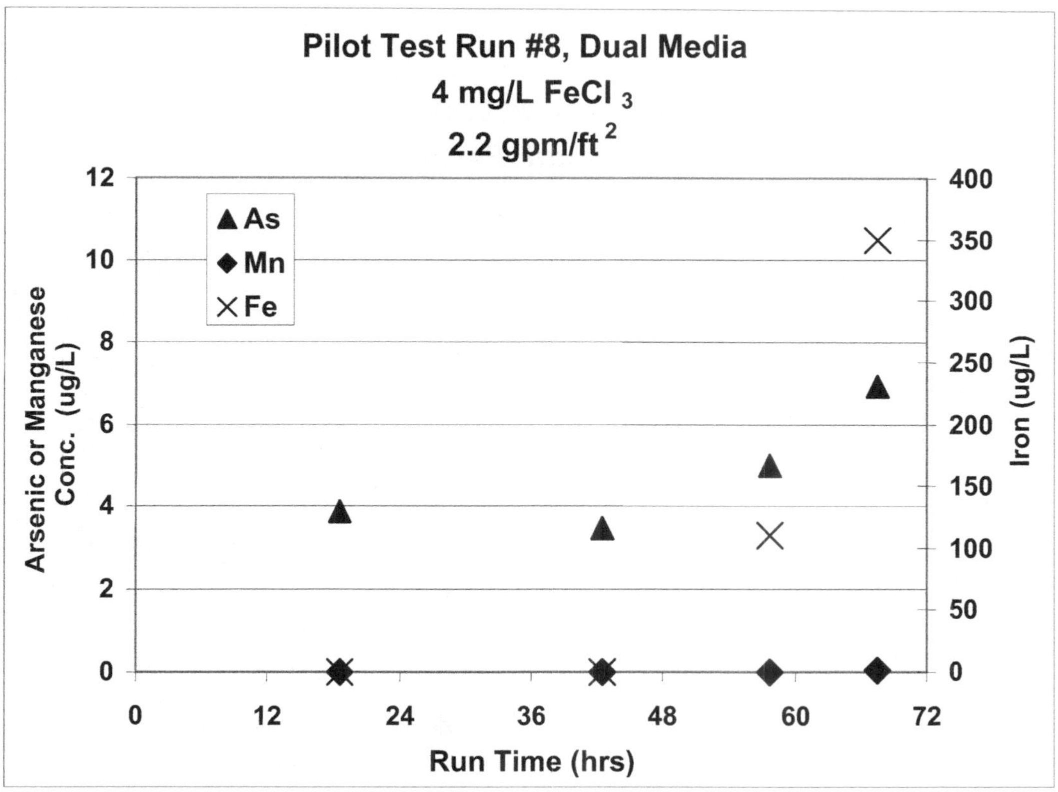

Figure B.2 Pilot test run #8, dual media, 4 mg/L $FeCl_3$, 2.2 gpm/ft^2

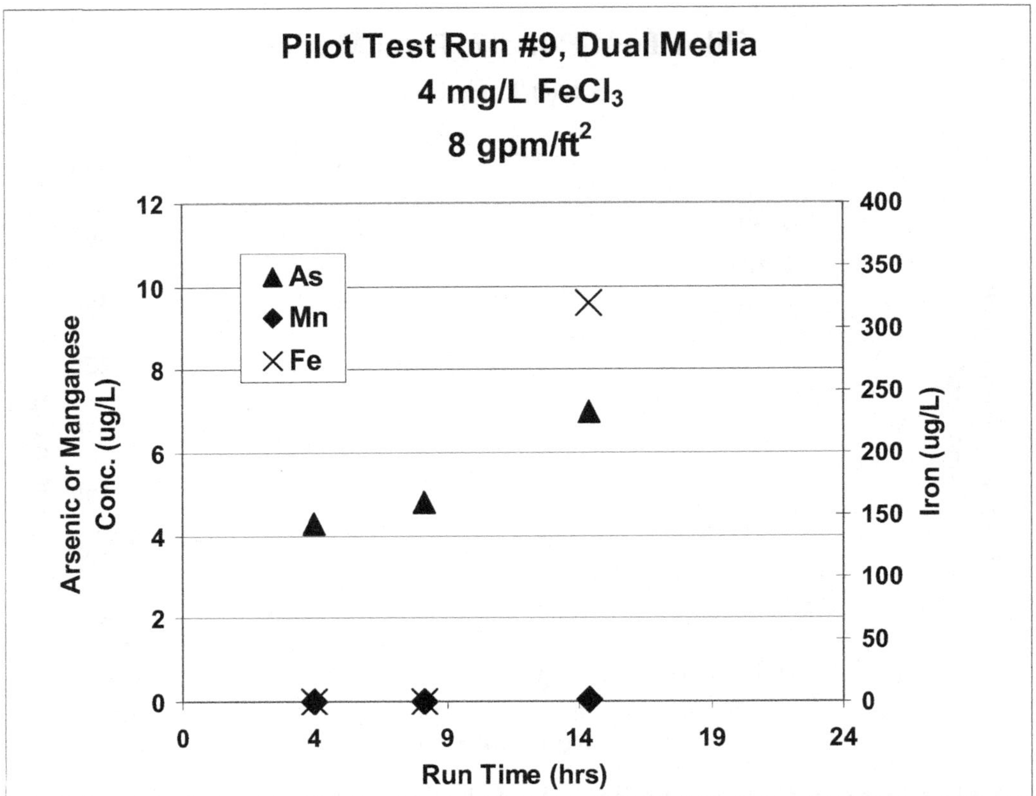

Figure B.3 Pilot test run #9, dual media, 4 mg/L $FeCl_3$, 8 gpm/ft^2

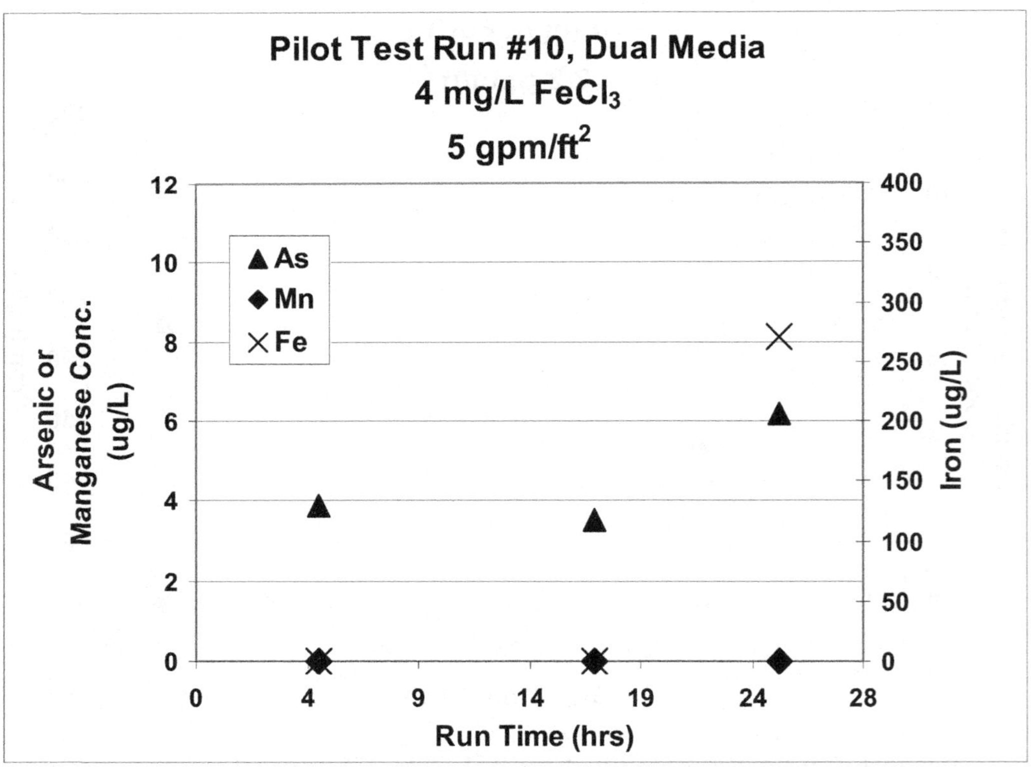

Figure B.4 Pilot test run #10, dual media, 4 mg/L $FeCl_3$, 5 gpm/ft^2

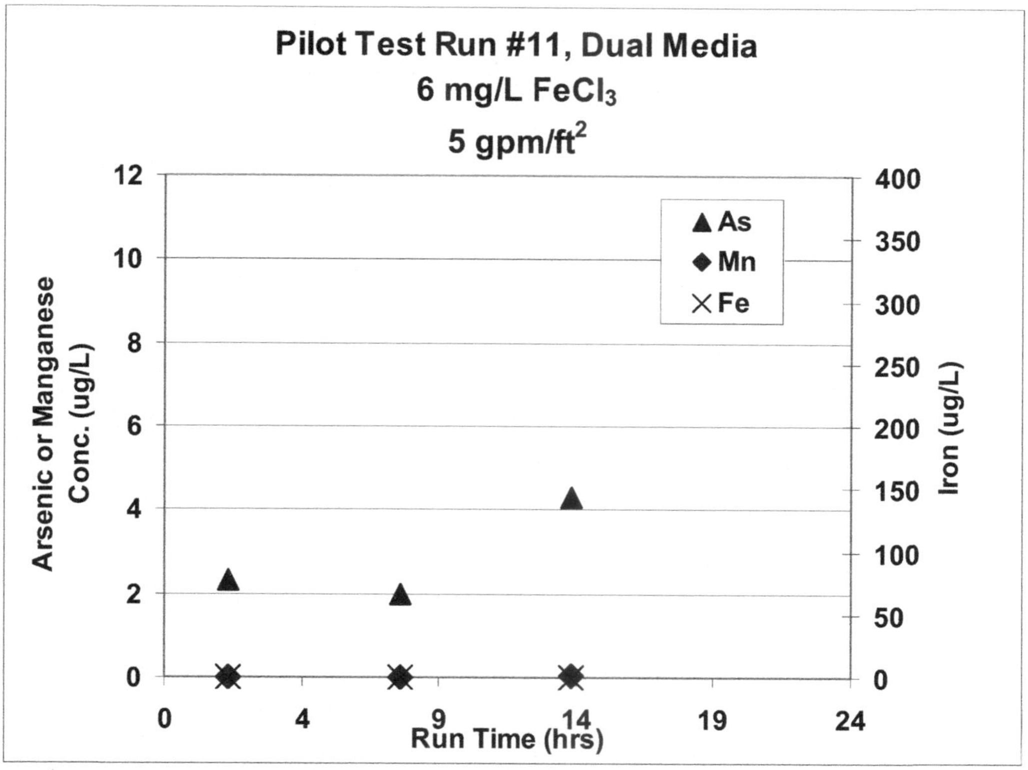

Figure B.5 Pilot test run #11, dual media, 6 mg/L $FeCl_3$, 5 gpm/ft^2

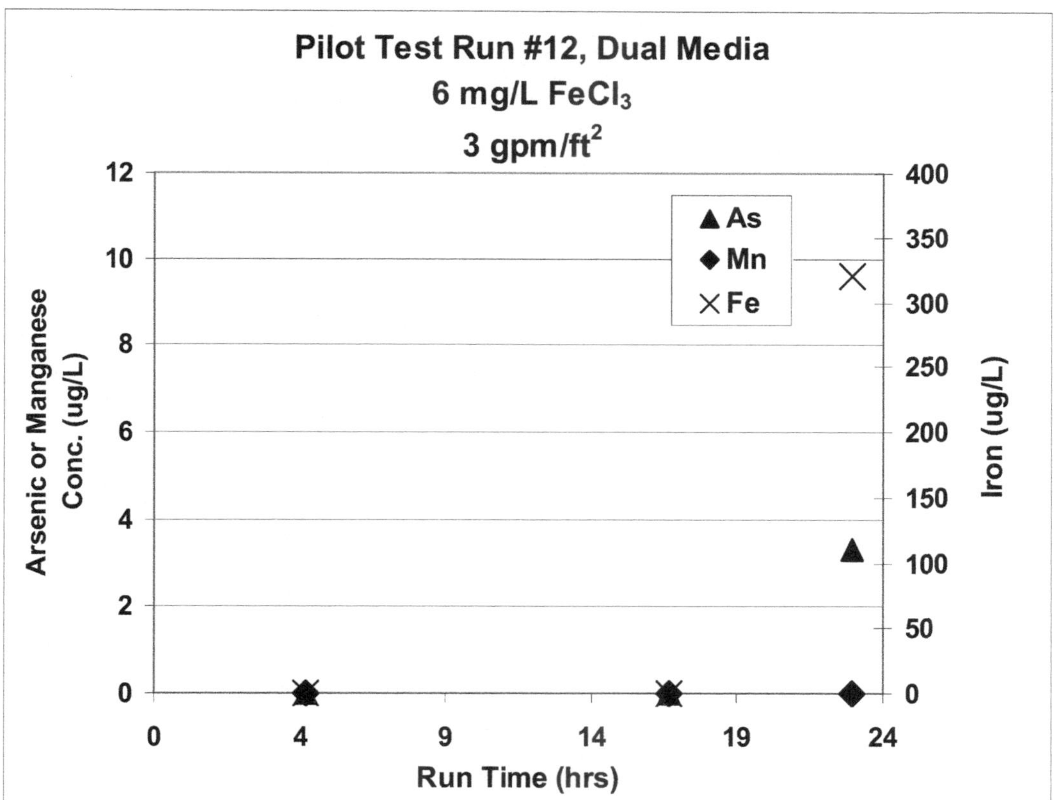

Figure B.6 Pilot test run #12, dual media, 6 mg/L $FeCl_3$, 3 gpm/ft^2

APPENDIX C
PHOTO LOG OF TEST SITE

Figure C.1 Mini-bulk tank for ferric chloride located near chlorine storage building

Figure C.2 Mini-bulk tank for ferric chloride located on concrete pad

Figure C.3 Chemical feed into chlorine storage unit

Figure C.4 Chemical feed into chlorine storage unit

Figure C.5 Chemical feed pump

Figure C.6 Chemical feed calibration cylinder

Figure C.7 Static mixer and chemical feed line located in vault

Figure C.8 Intake water piping and control building

Figure C.9 Pressure Filters

Figure C.10 De-chlorination port for backwash water

Figure C.11 Backwash holding tanks, piping, and automated decant valve

Figure C.12 Sludge thickening tank

APPENDIX D
FULL-SCALE TEST RESULTS

Table D.1
Summary of full scale test runs

Run Number	Start Date	Dose (mg/L FeCl3)	Break-through Reached? (y/n)	Chart Location
1	5/2/2002	2	n	D-3
2	5/6/2002	2	n	D-3
3	6/10/2002	2	y	D-4
4	6/17/2002	2	n	D-4
5	6/19/2002	2	n	D-5
6	6/21/2002	2	n	D-5
7	6/24/2002	4	y	D-6
9	7/8/2002	6	y	D-6
10	7/10/2002	8	y	D-7
11	7/16/2002	10	y	D-7
12	7/17/2002	10	y	D-8
14	7/18/2002	14	y	D-8
15	7/19/2002	2	n	D-9
16	8/1/2002	4	n	D-9
17	8/5/2002	4	n	D-10
18	8/6/2002	4	n	D-10
19	8/7/2002	6	n	D-11
20	8/8/2002	6	n	D-11
21	8/12/2002	6	y	D-12
22	8/14/2002	8	y	D-12
23	8/15/2002	8	n	D-13
24	8/19/2002	4	n	D-13
25	8/21/2002	4	n	D-14
26	8/26/2002	4	n	D-14
27	8/27/2002	6	n	D-15
28	8/28/2002	6	n	D-15

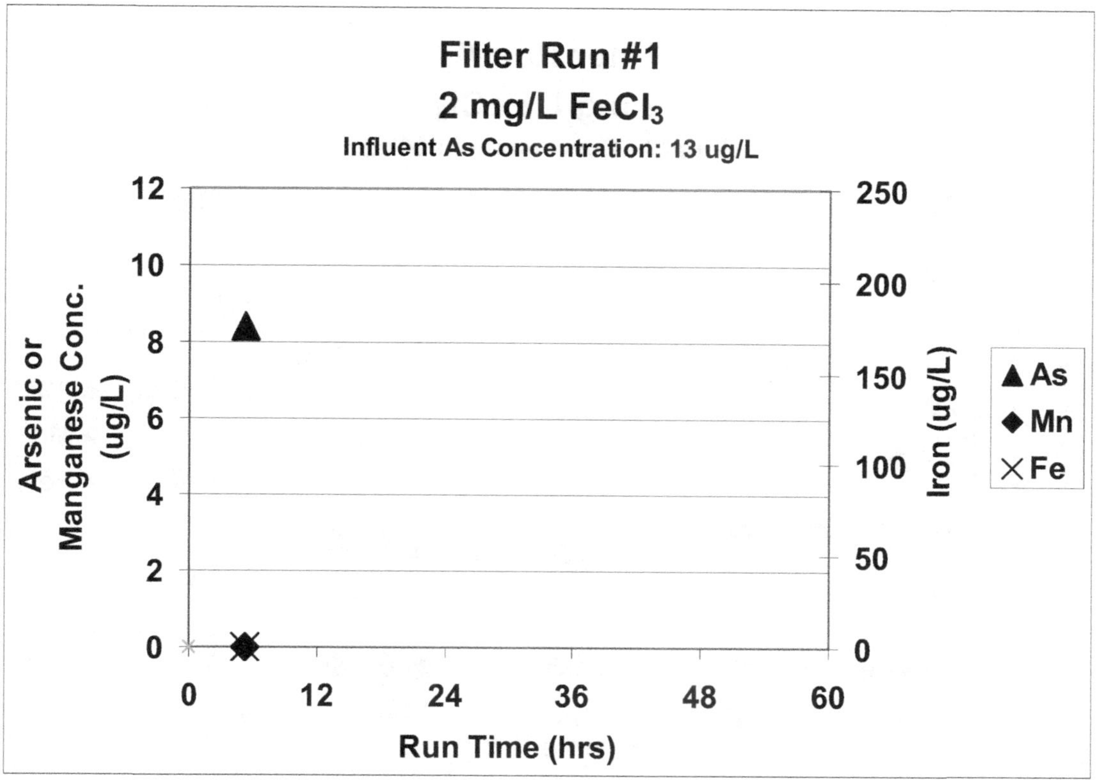

Figure D.1 Filter run #1, 2 mg/L $FeCl_3$, influent As concentration 13 μg/L

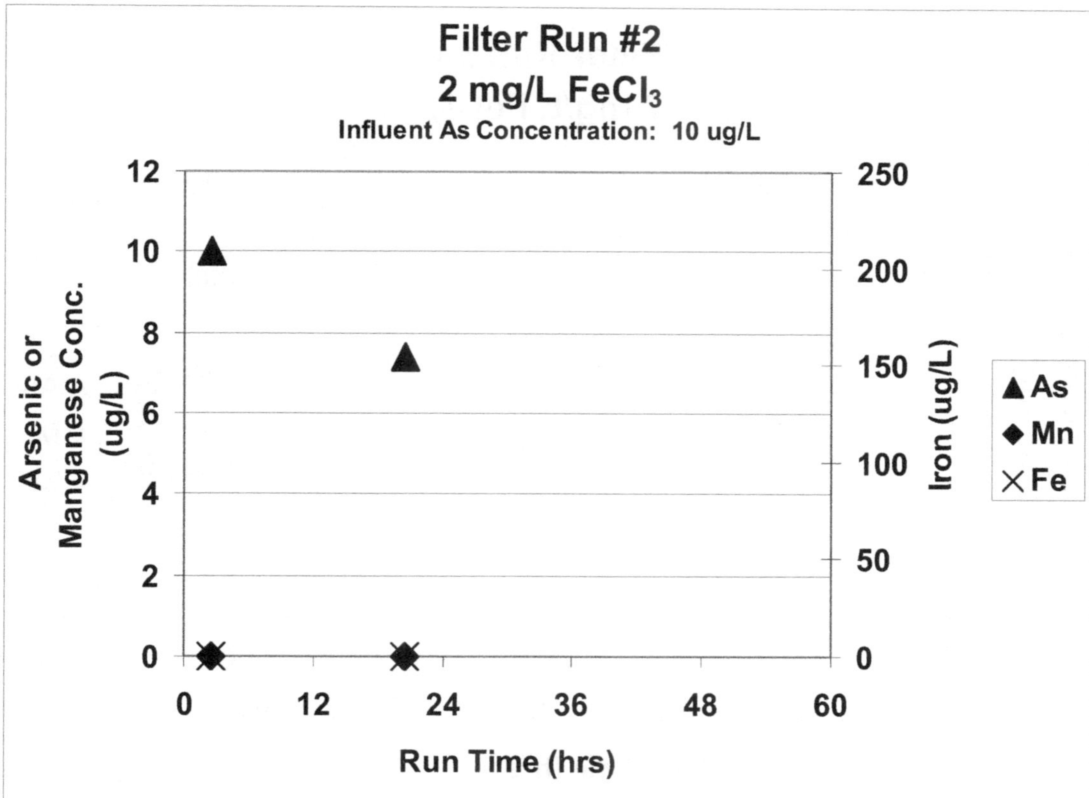

Figure D.2 Filter run #2, 2 mg/L $FeCl_3$, influent As concentration 10 μg/L

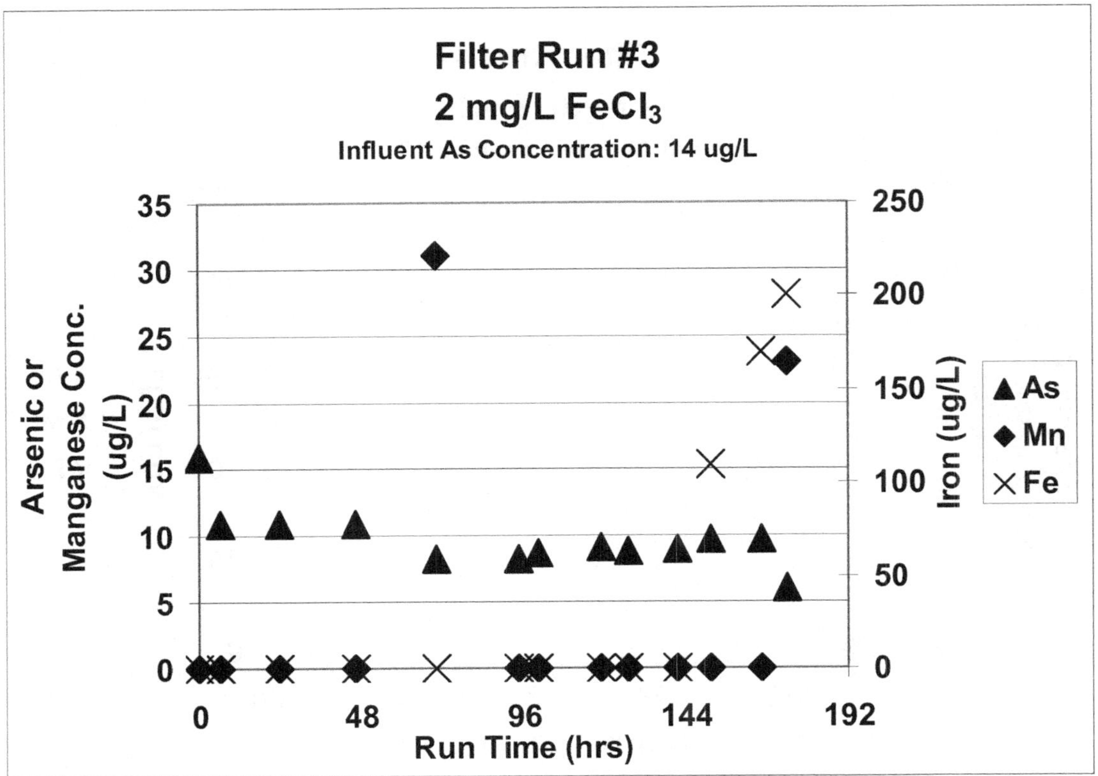

Figure D.3 Filter run #3, 2 mg/L $FeCl_3$, influent As concentration 14 µg/L

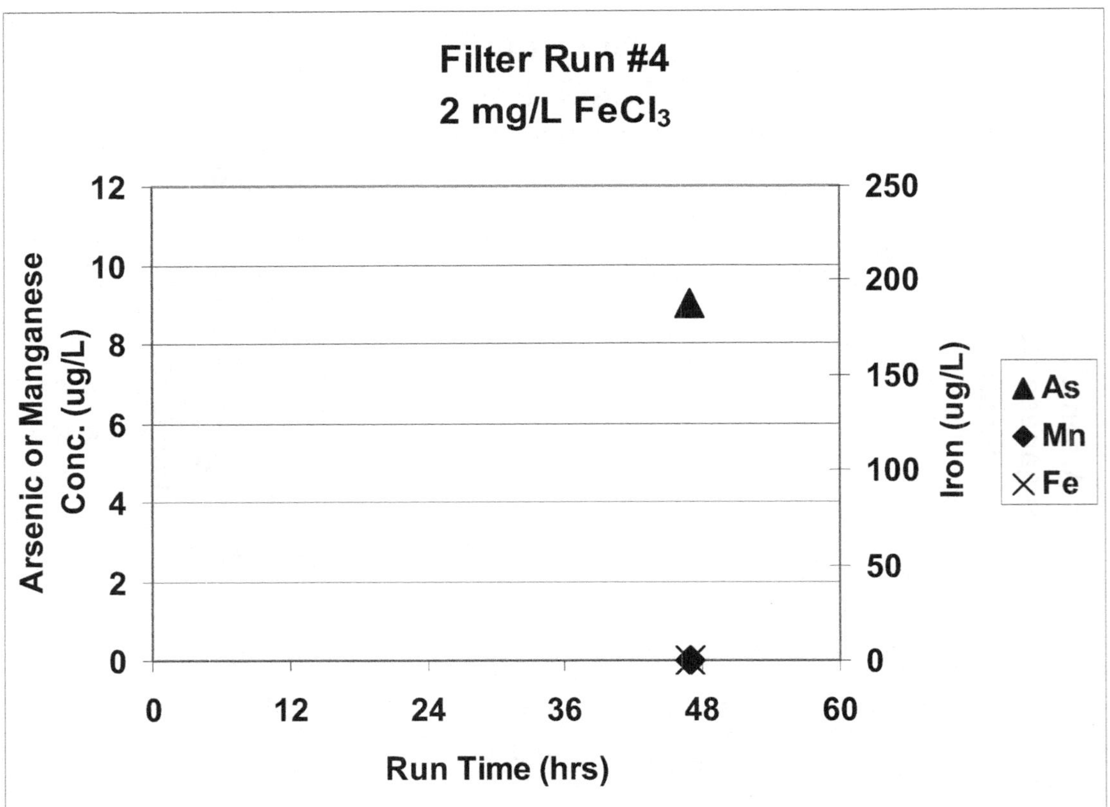

Figure D.4 Filter run #4, 2 mg/L $FeCl_3$

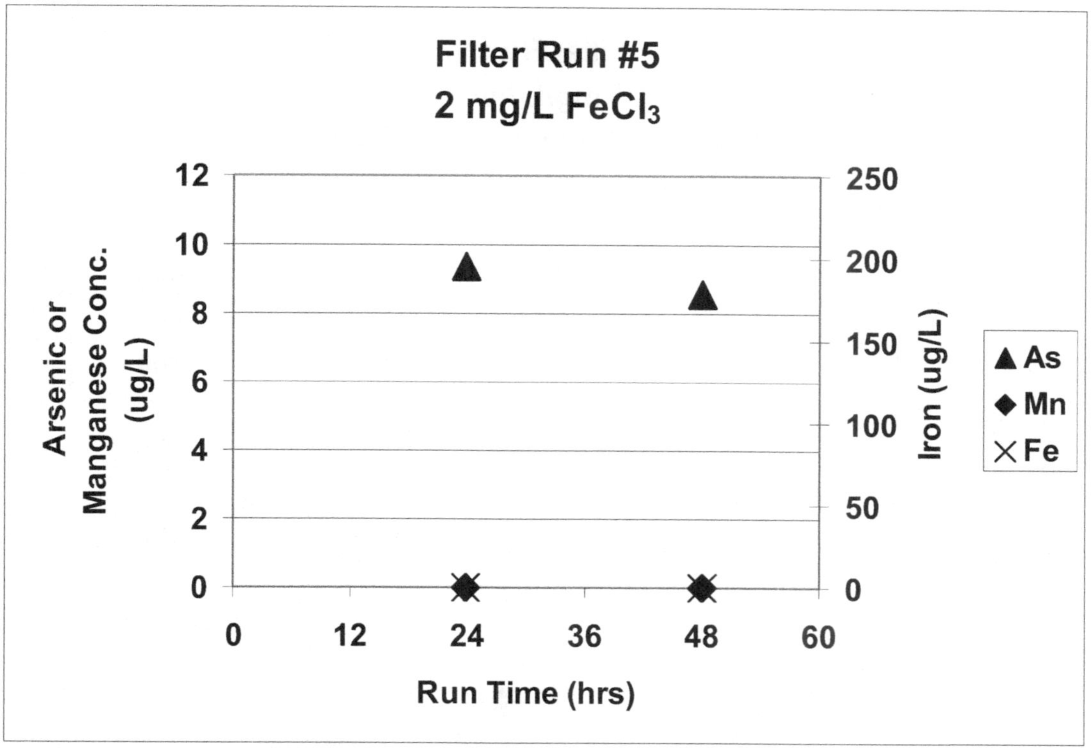

Figure D.5 Filter run #5, 2 mg/L $FeCl_3$

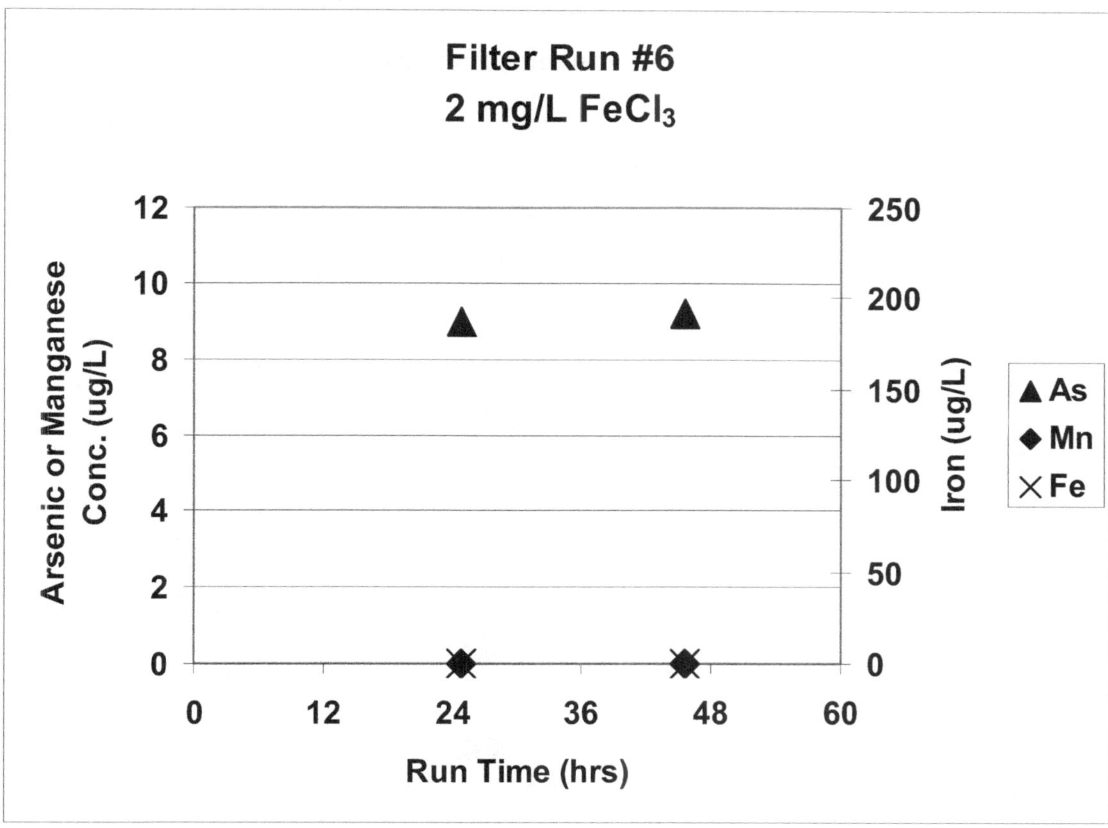

Figure D.6 Filter run #6, 2 mg/L $FeCl_3$

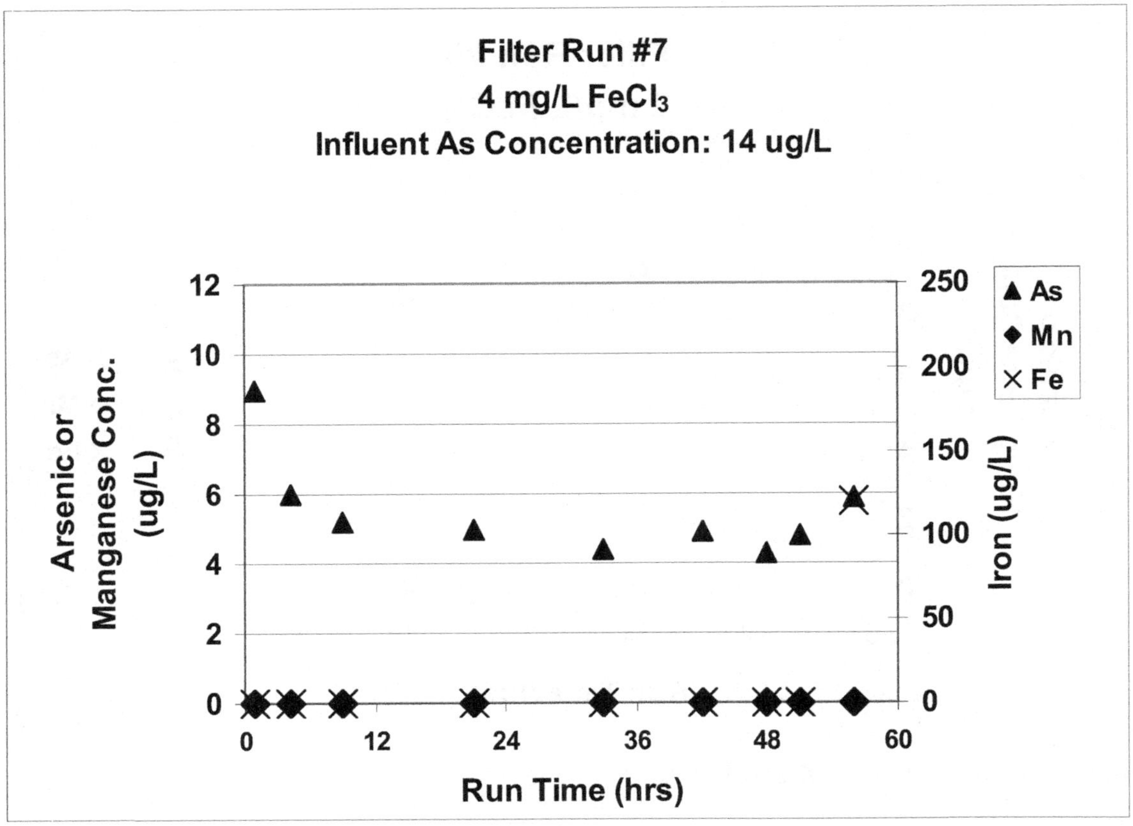

Figure D.7 Filter run #7, 4 mg/L $FeCl_3$, influent As concentration 14 μg/L

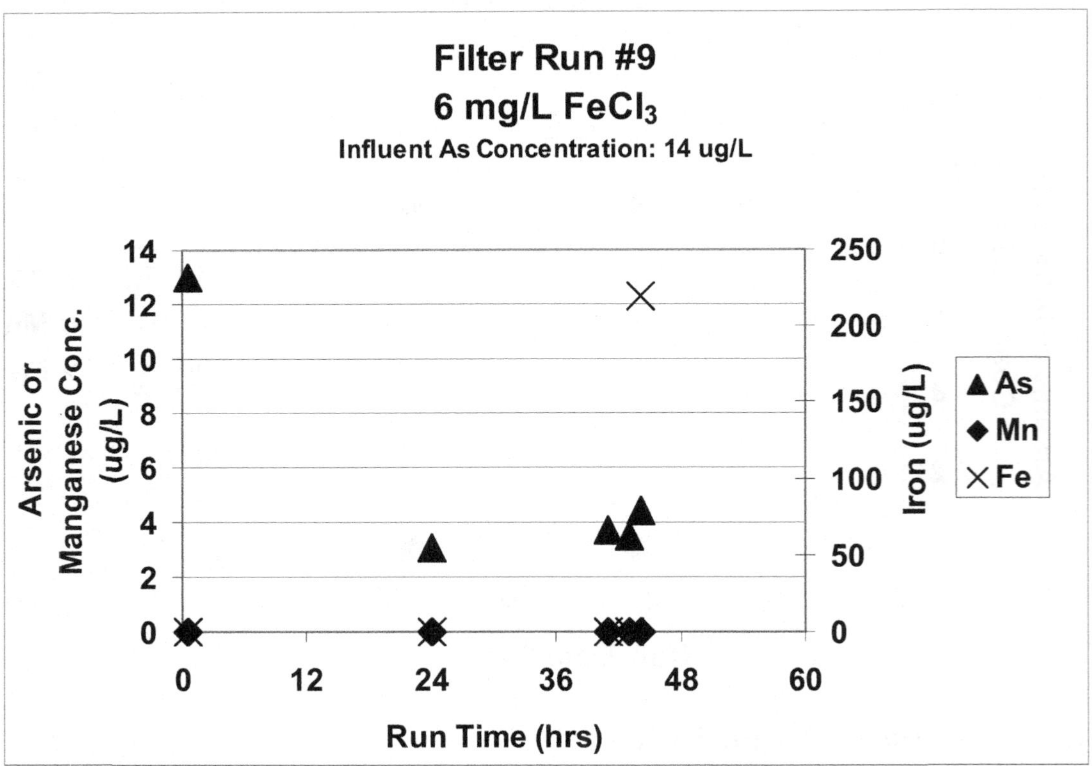

Figure D.8 Filter run #9, 6 mg/L $FeCl_3$, influent As concentration 14 μg/L

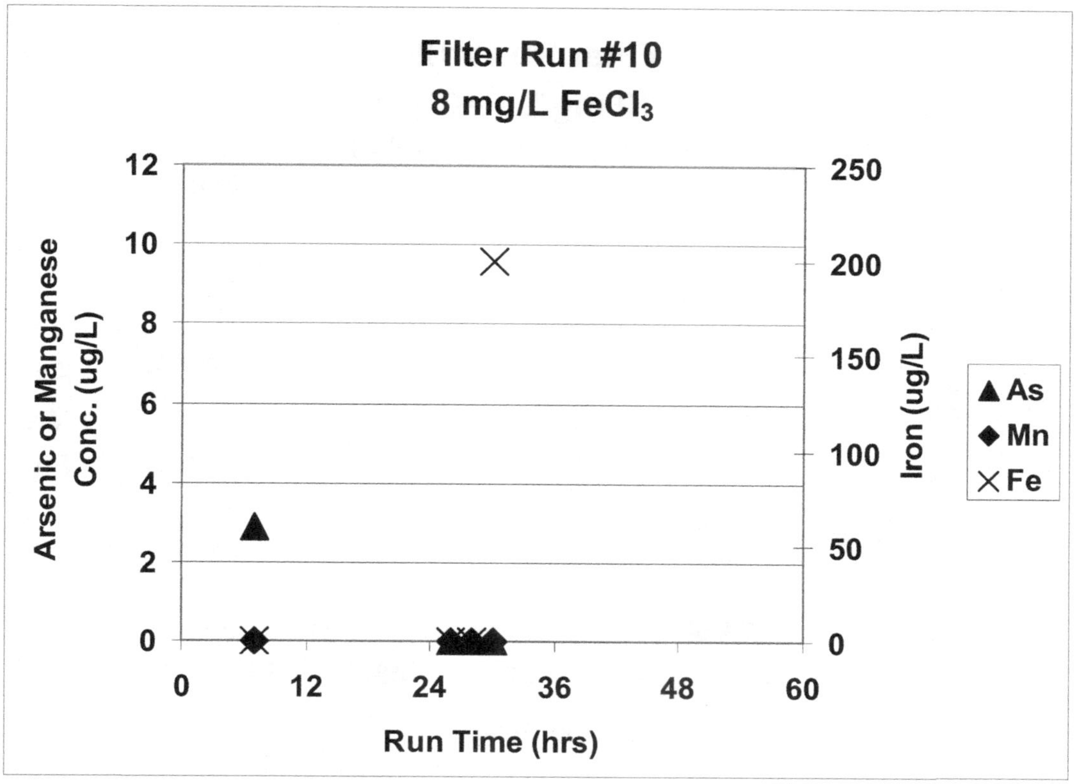

Figure D.9 Filter run #10, 8 mg/L $FeCl_3$

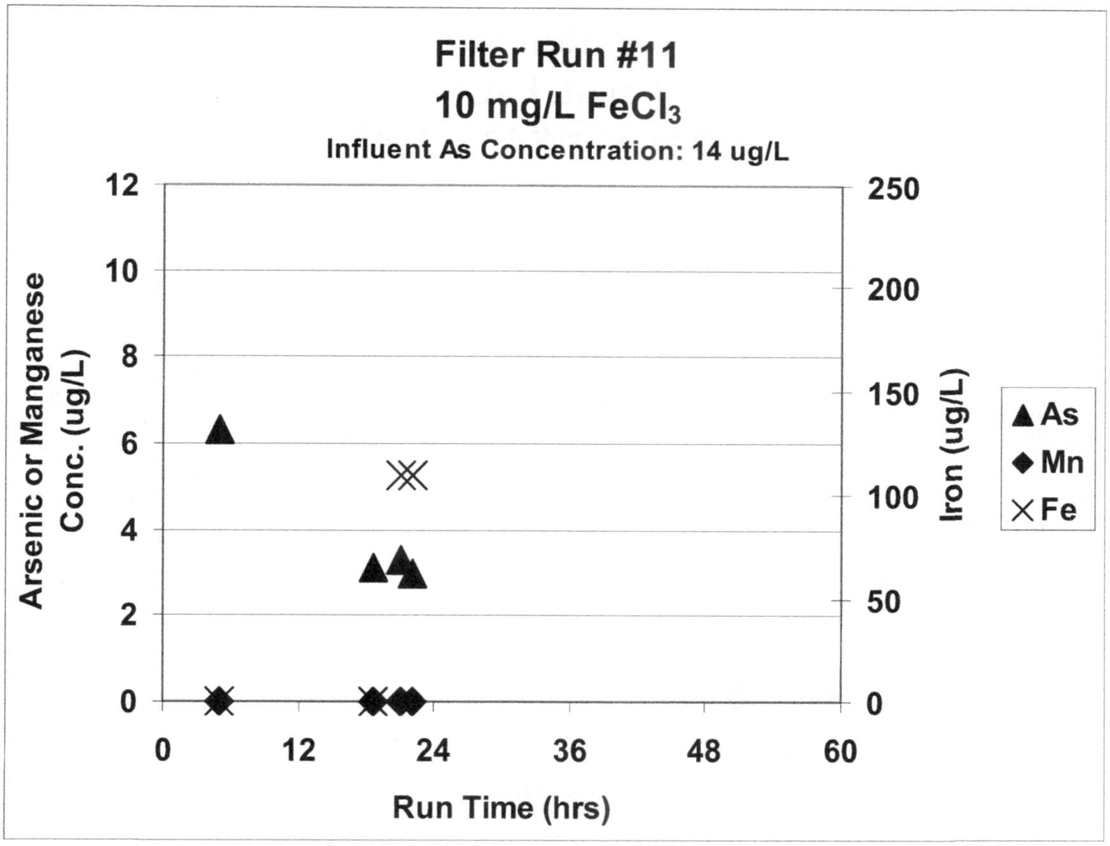

Figure D.10 Filter run #11, 10 mg/L $FeCl_3$, influent As concentration 14 μg/L

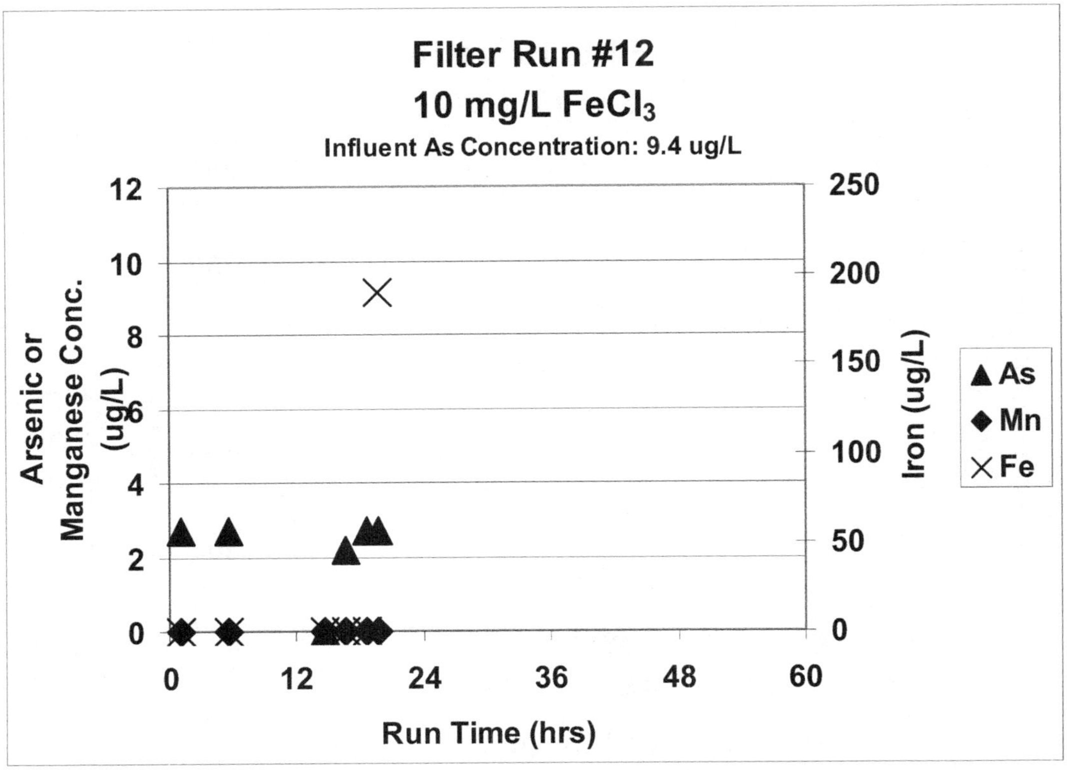

Figure D.11 Filter run #12, 10 mg/L $FeCl_3$, influent As concentration 9.4 μg/L

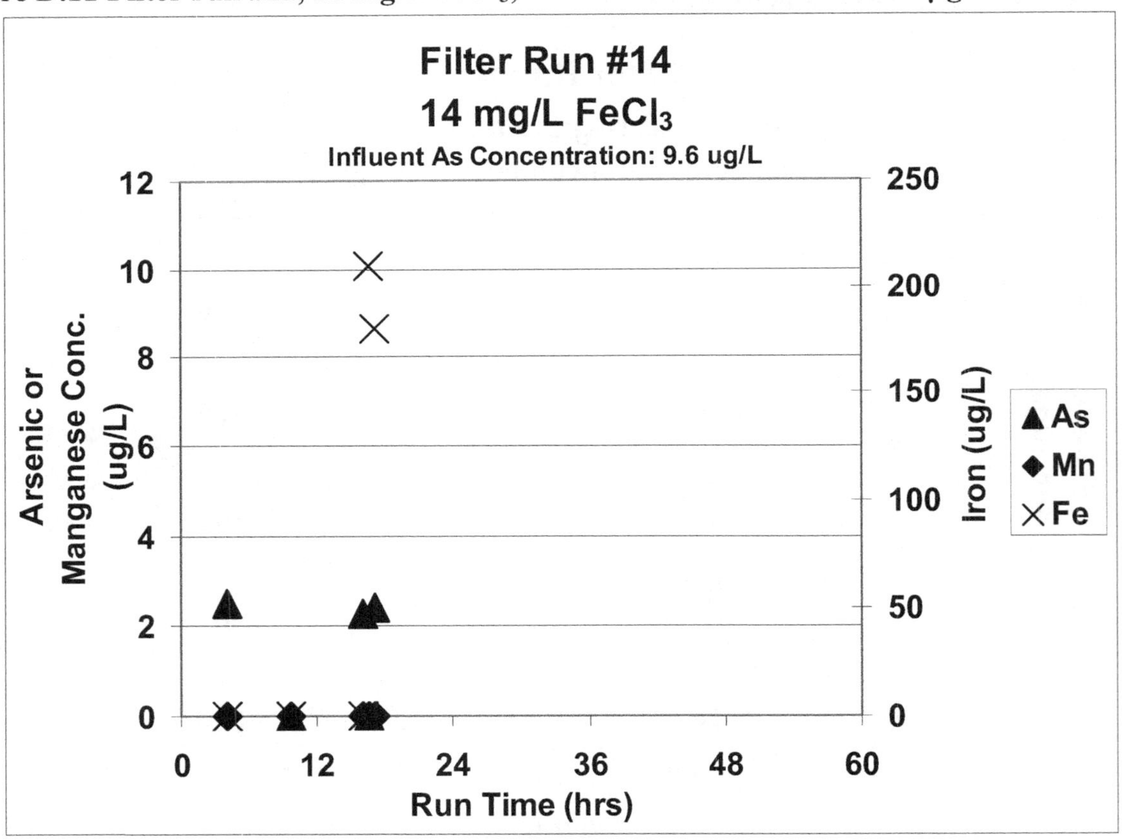

Figure D.12 Filter run #14, 14 mg/L $FeCl_3$, influent As concentration 9.6 μg/L

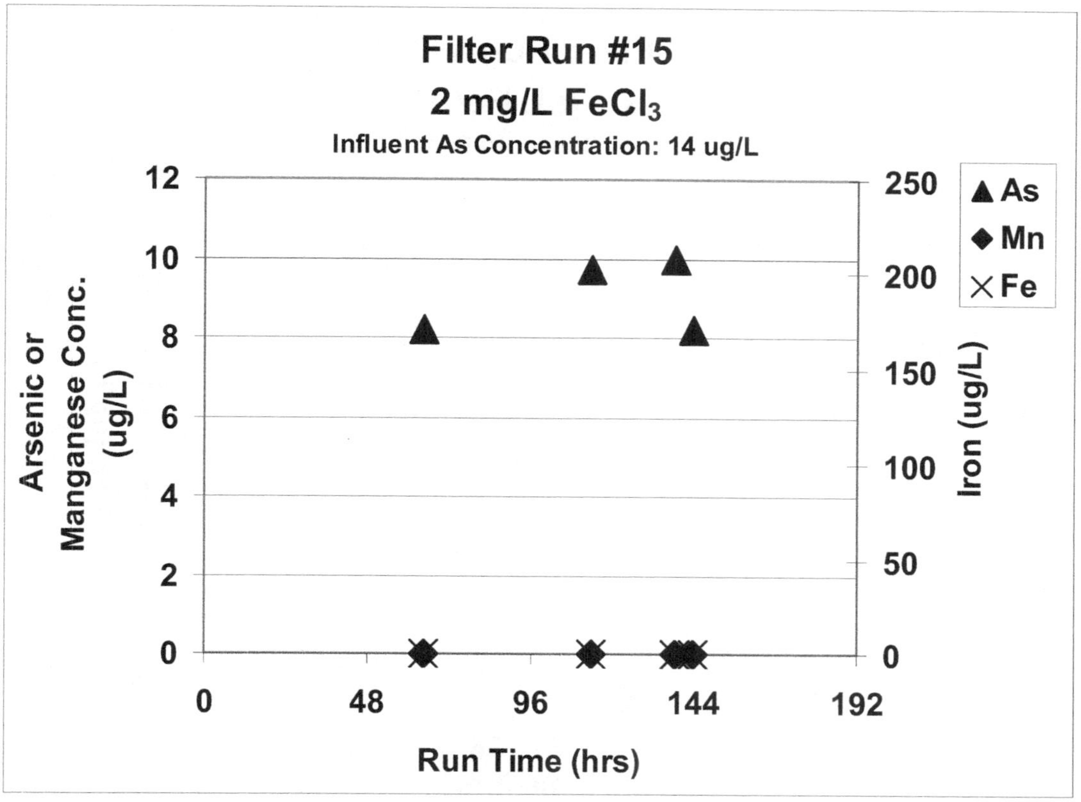

Figure D.13 Filter run #15, 2 mg/L $FeCl_3$, influent As concentration 14 µg/L

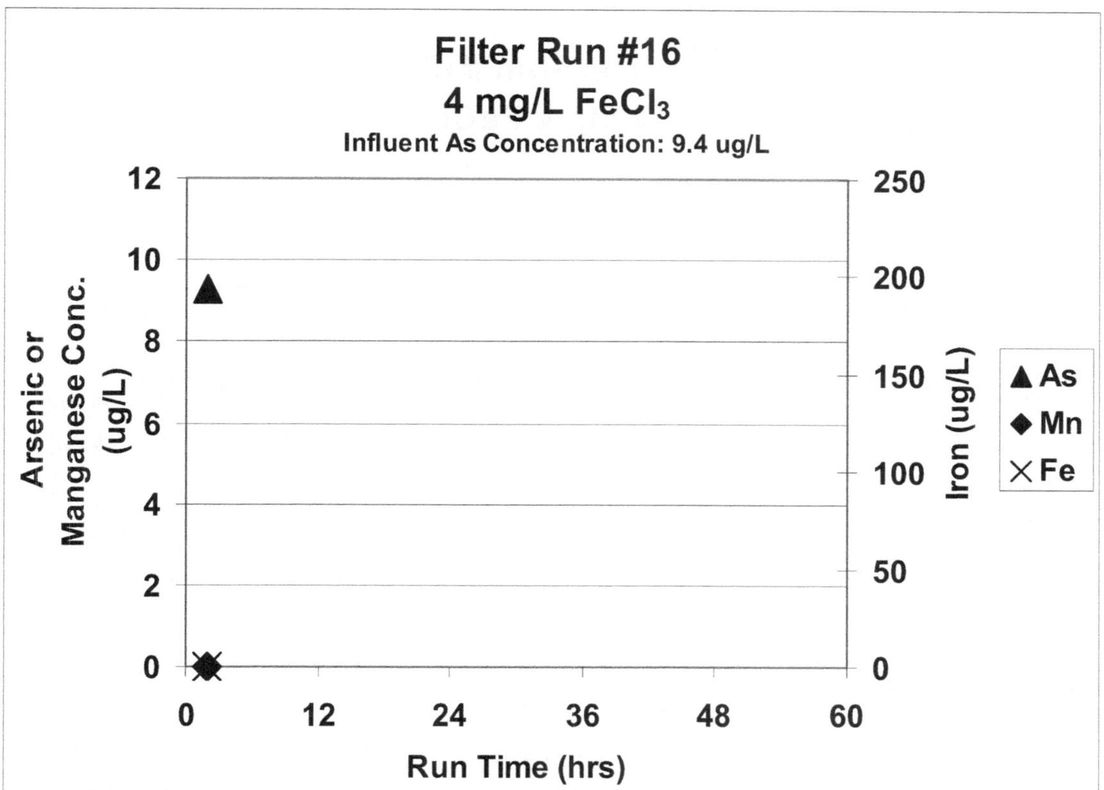

Figure D.14 Filter run #16, 4 mg/L $FeCl_3$, influent As concentration 9.4 µg/L

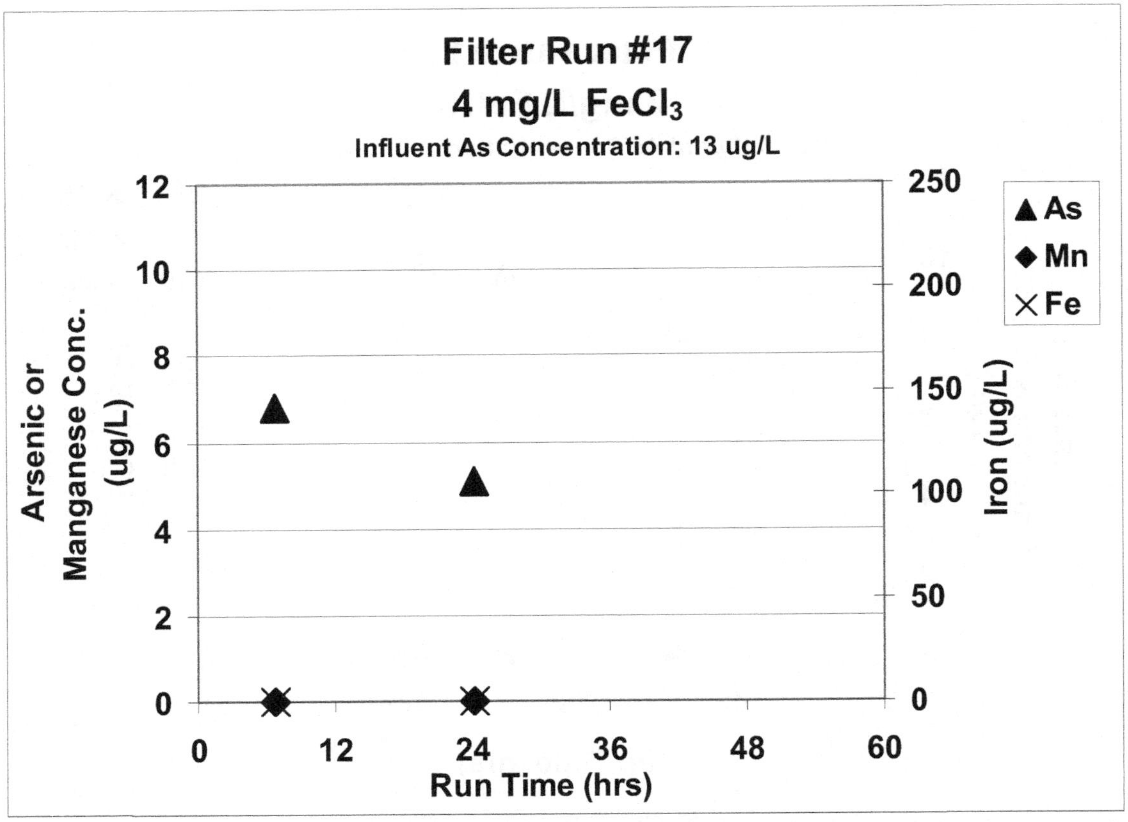

Figure D.15 Filter run #17, 4 mg/L $FeCl_3$, influent As concentration 13 μg/L

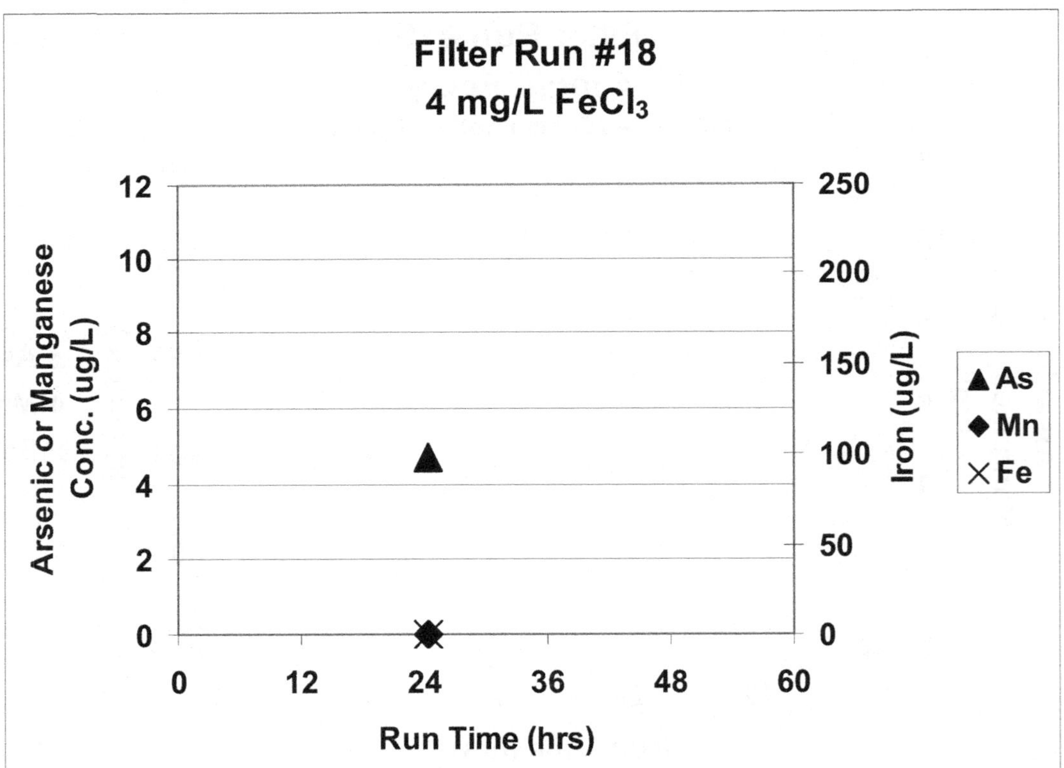

Figure D.16 Filter run #18, 4 mg/L $FeCl_3$

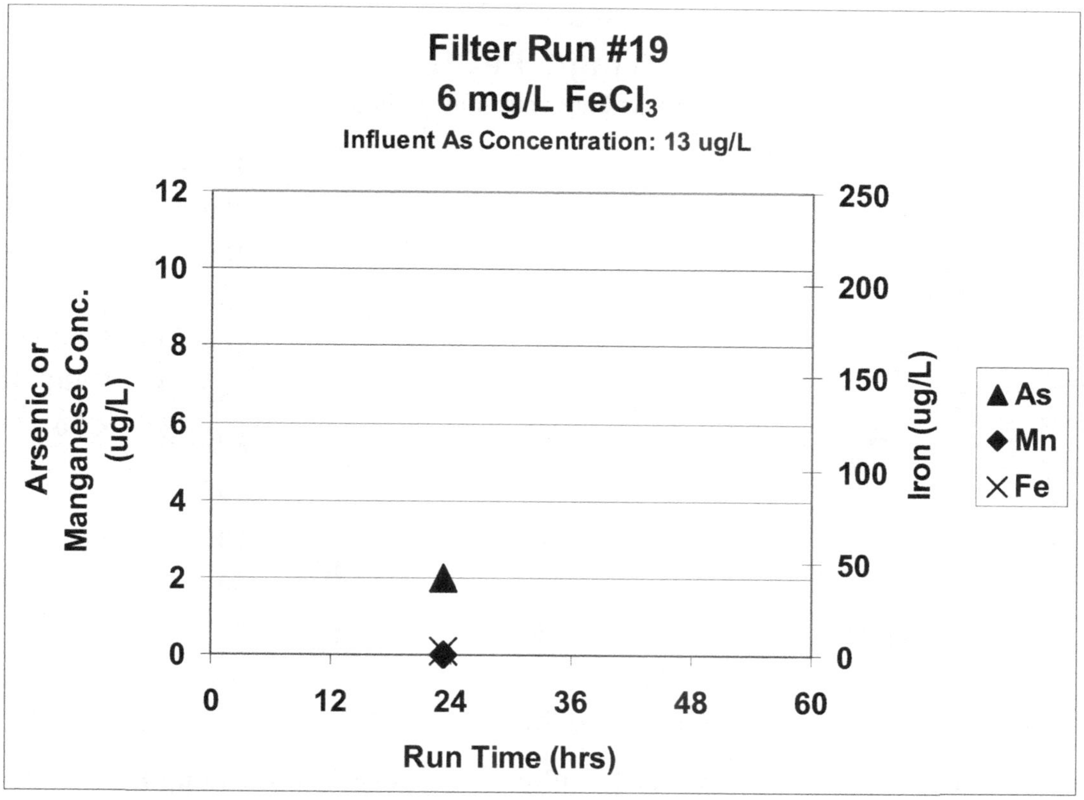

Figure D.17 Filter run #19, 6 mg/L $FeCl_3$, influent As concentration 13 μg/L

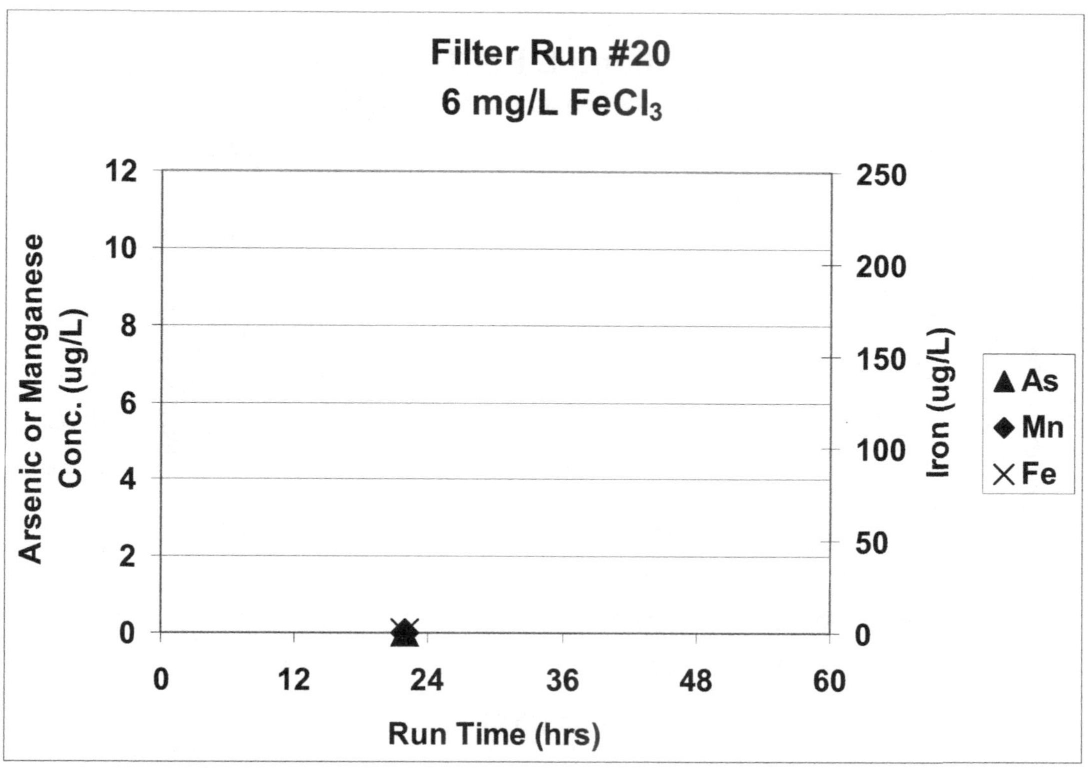

Figure D.18 Filter run #20, 6 mg/L $FeCl_3$

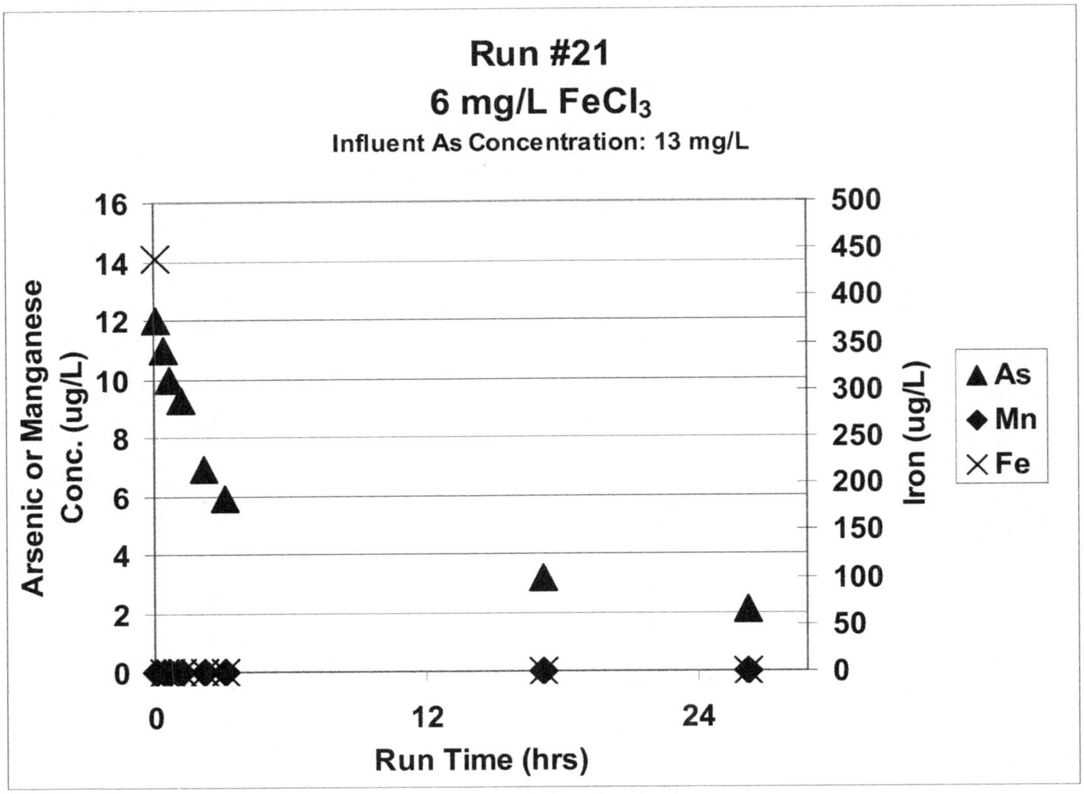

Figure D.19 Filter run #21, 6 mg/L $FeCl_3$, influent As concentration 13 μg/L

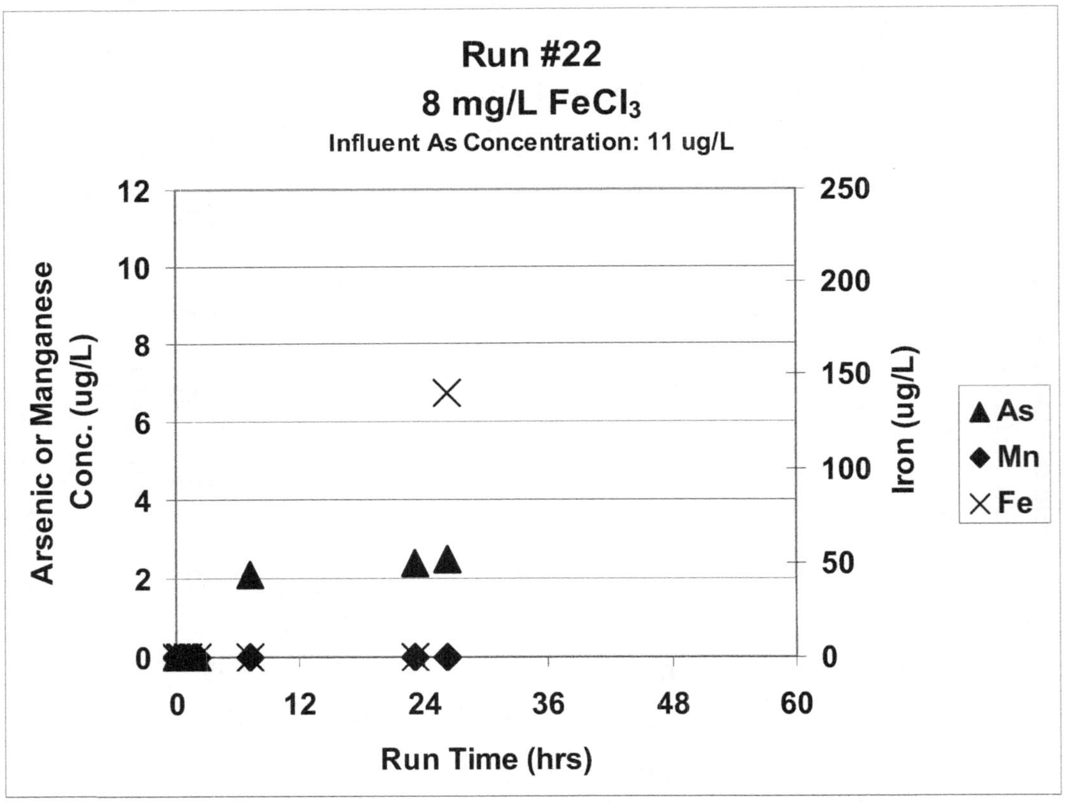

Figure D.20 Filter run #22, 8 mg/L $FeCl_3$, influent As concentration 11 μg/L

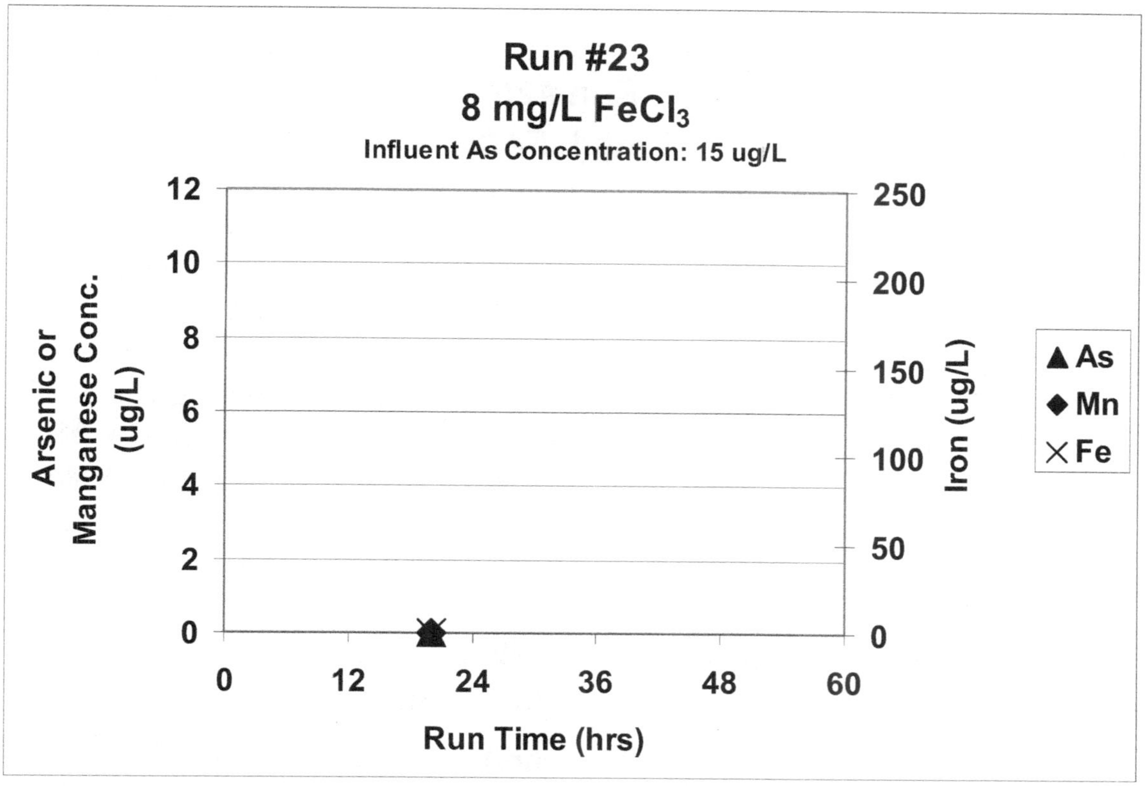

Figure D.21 Filter run #23, 8 mg/L $FeCl_3$, influent As concentration 15 μg/L

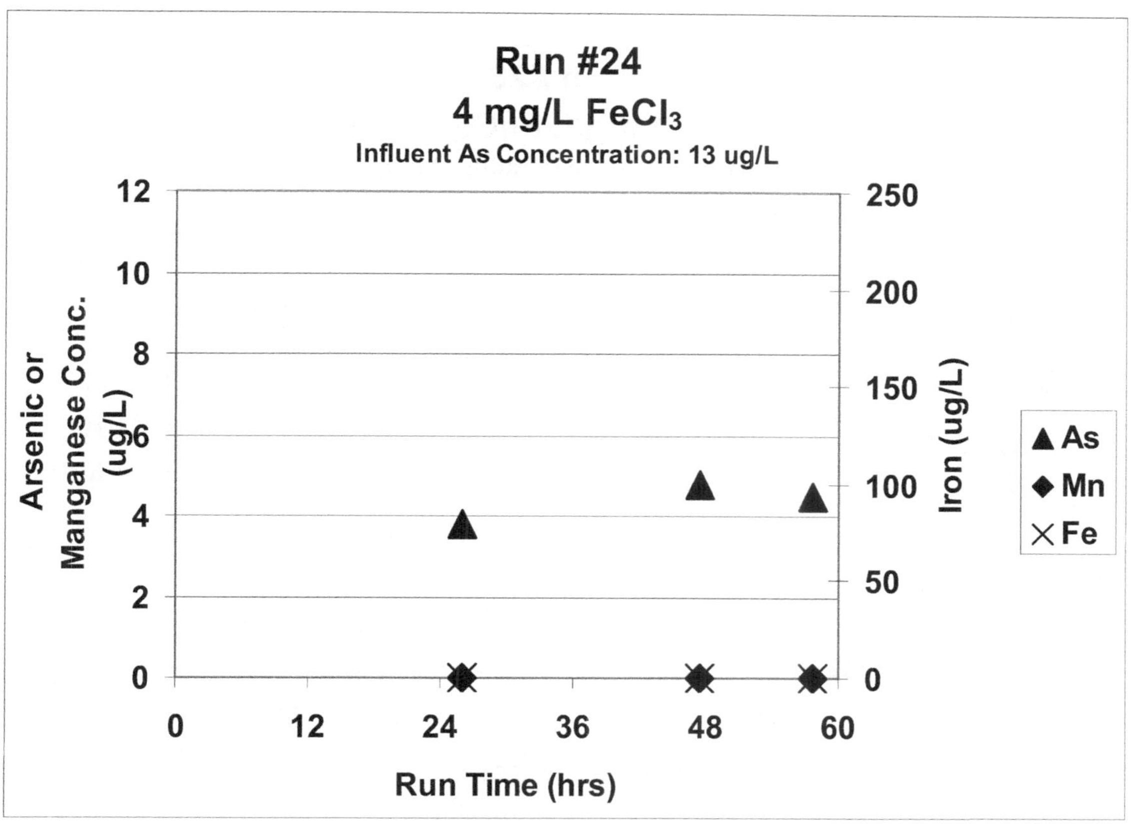

Figure D.22 Filter run #24, 4 mg/L $FeCl_3$, influent As concentration 13 μg/L

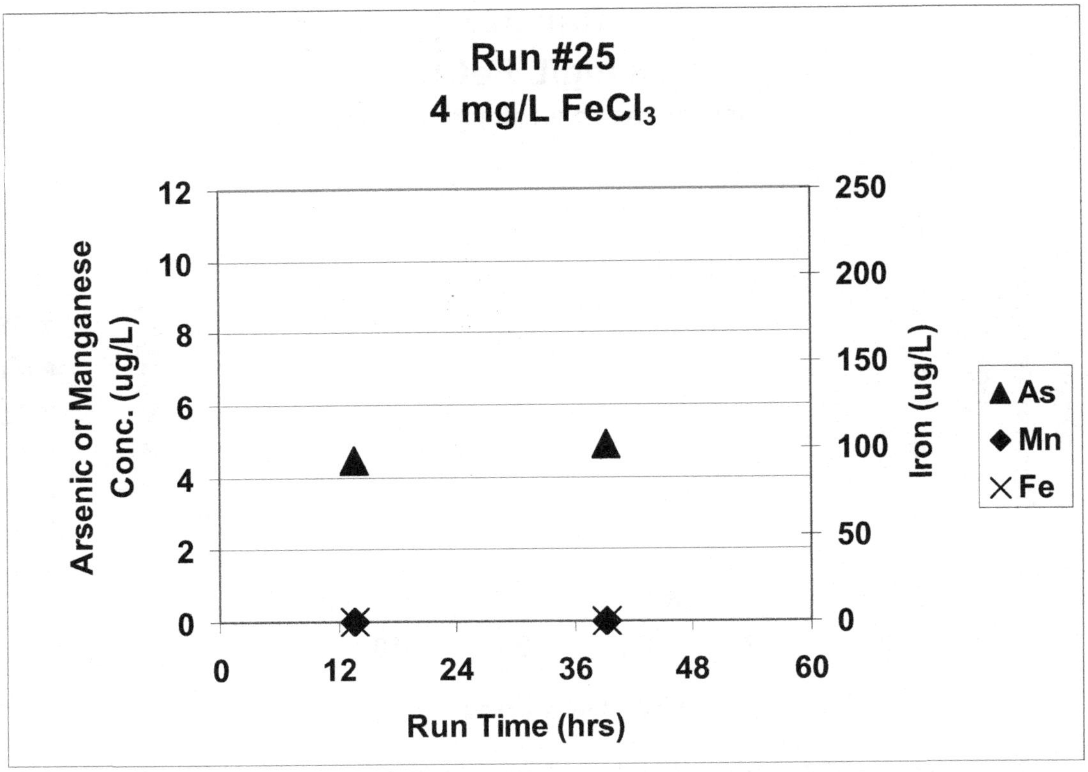

Figure D.23 Filter run #25, 4 mg/L $FeCl_3$

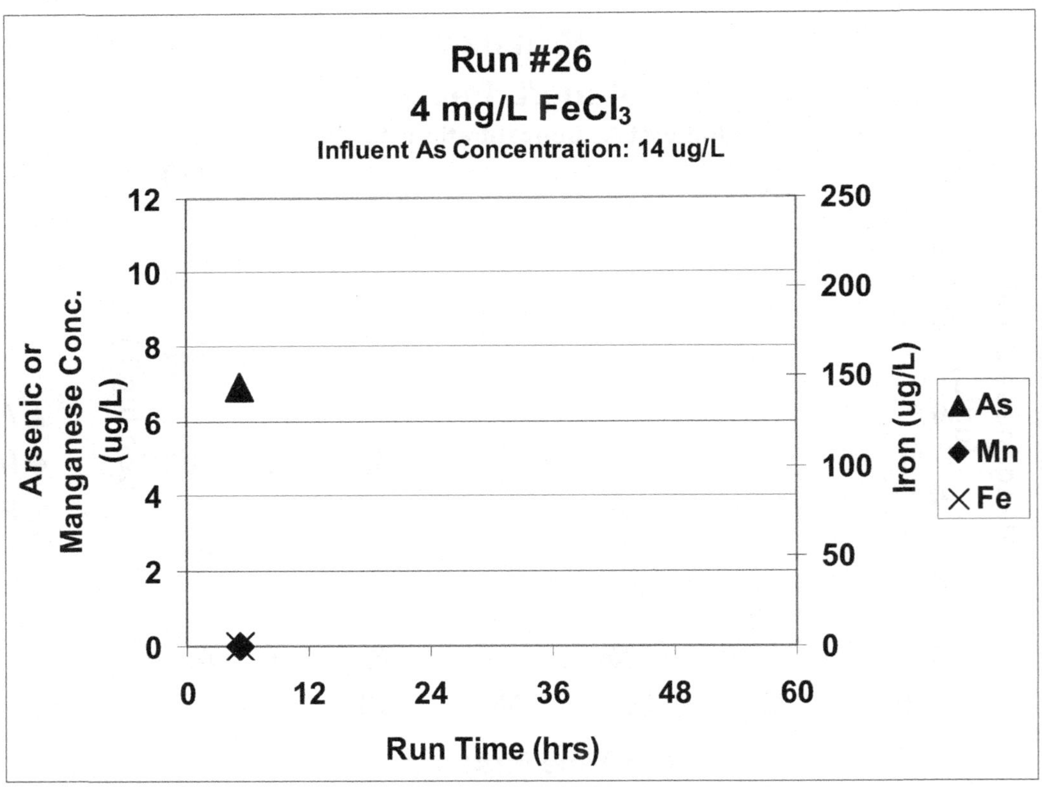

Figure D.24 Filter run #26, 4 mg/L $FeCl_3$, influent As concentration 14 μg/L

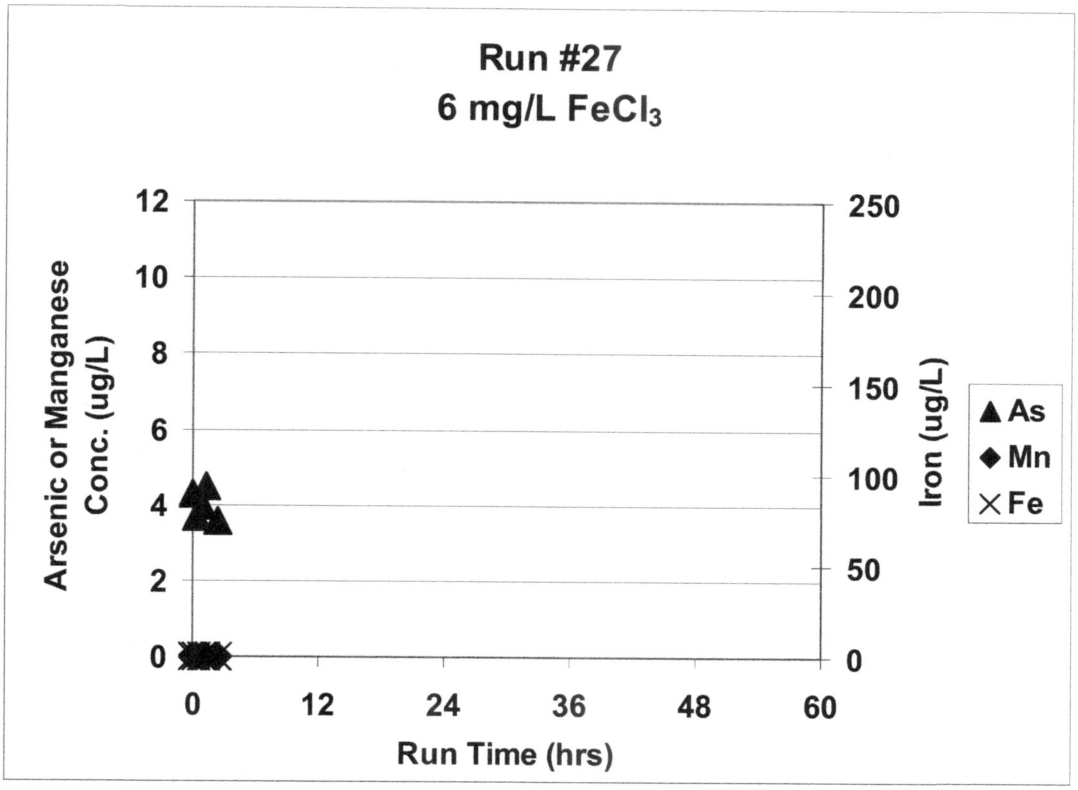

Figure D.25 Filter run #27, 6 mg/L $FeCl_3$

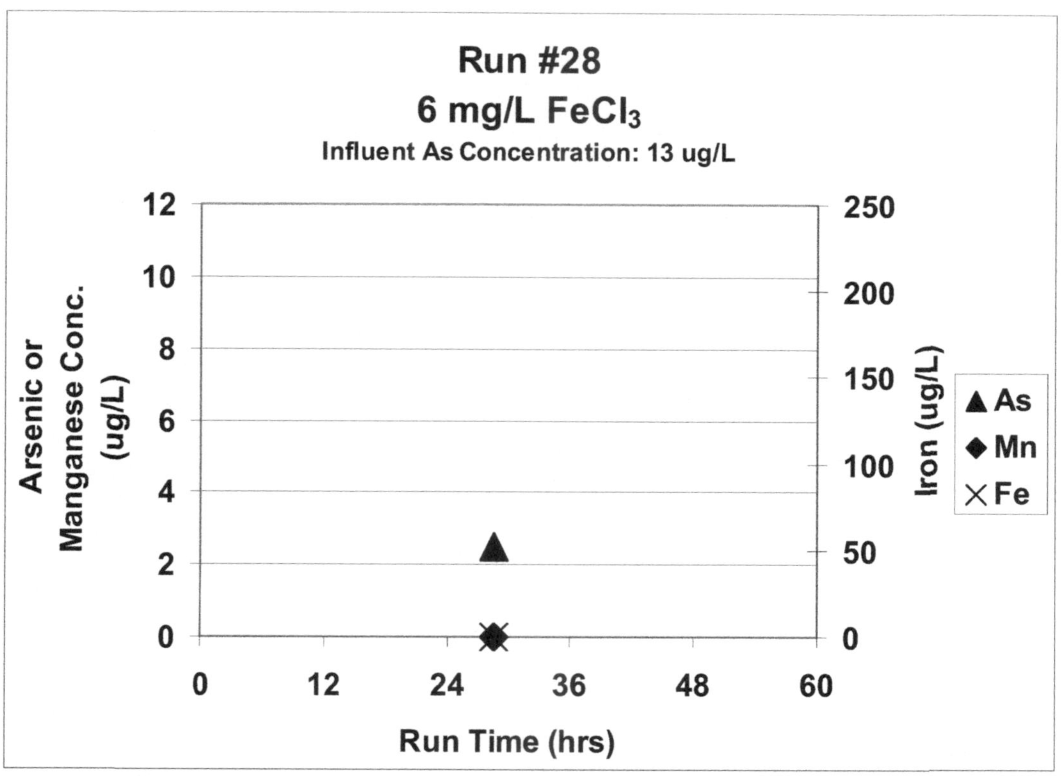

Figure D.26 Filter run #28, 6 mg/L $FeCl_3$, influent As concentration 13 μg/L

APPENDIX E
COST ANALYSIS

Table E.1
Detailed cost analysis for 2.7 mgd plant capacity

Lakewood, CA Created: 1/4/2004

Arsenic and Manganese Removal Facility Revision: 1

Arsenic Removal - Ferric Chloride and Pressure Filter- Plant Capacity -2.7 MGD

Item	Description	Units	Quant.	Unit Cost	Subtotal	Total Cost Yr 2004 Dollars	
	Construction Cost Estimate						
1	**General Administration**						
1.01	Mobilization/Bond/Insurance	ls	1	85,000	85,000		
1.02	Environmental Protection	ls	1	15,000	15,000		
1.03	Plant Startup	ls	1	30,000	30,000		
1.04	Record Drawings	ls	1	8,500	8,500		
			Subtotal General Administration			**$138,500**	
2	**Yard, Piping, Site Grading, and Restoration**						
2.01	Site Preparation	ls	1	6,000	6,000		
2.02	Yard Piping	ls	400	45	18,000		
2.03	Earthwork	ls	1	11,500	11,500		
2.04	Paving	sy	1200	29	34,800		
2.05	Restoration and Landscaping	ls	1	11,500	11,500		
			Yard Piping, Site Grading, and Restoration			**$81,800**	
3	**Treatment Plant Building**						
3.01	Plumbing Allowance	ls	1	11,500	11,500		
3.02	Process Piping	ls	1	30,000	30,000		
3.03	Static Mixer	ls	2	5,000	10,000		
3.04	Backwash Drain Piping	ls	1	17,500	17,500		
3.05	Pressure Filters/Media/Automated Valves	ea	4	120,000	480,000		
3.06	Finishing	ls	1	29,000	29,000		
3.07	Grating	ls	1	6,000	6,000		
3.08	Misc. Structures	ls	1	6,000	6,000		
3.09	Misc. Metals	ls	1	6,000	6,000		
3.10	Instrumentation and Controls	ls	1	75,000	75,000		
3.11	Telementry Programming	ls	1	75,000	75,000		
3.12	Electrical Transformer	ls	1	145,000	145,000		365,000
3.13	Standby Power	ls	1	70,000	70,000		
3.14	HVAC Allowance	sf	2500	5	12,500		20.0%

3.15	Building	sf	2500	105	262,500		
		Subtotal Treatment Plant Building				**$1,236,000**	
4	**Chem Feed and Backwash System**						
4.01	Ferric Chloride Feed System	ls	1	30,000	30,000		
4.02	Polymer Feed System	ls	1	25,000	25,000		
4.03	Hypochlorite Feed System	ls	1	30,000	30,000		
4.04	Chemical Feed Piping	ls	3	6,000	18,000		
4.05	Lab and Office Equipment	ls	1	9,500	9,500		
4.06	Recycle Pump Sta	ls	1	50,000	50,000		
4.07	Backwash Storage Basin	ea	2	55,000	110,000		
4.08	Backwash Blowers and Piping	ea	2	14,000	28,000		
4.09	Electrical & Controls	ls	1	70,000	70,000		
	Subtotal Backwash Modifications and Equipment					**$370,500**	
					Construction Subtotal	**$1,826,800**	
		Contingency		30%	548,000		
					Estimated Construction Costs	**$2,375,000**	
	Estimate Indirect Costs						
	Geotechnical Engineering				20,000		
	Surveying				30,000		
	Property Acquisition (Acres)			1	60,000		
	Engineering Report				25,000		
	Design Engineering			14%	332,500		
	Construction Services and Startup			7%	166,300		
	Administration			2%	47,500		
	Legal and Permits			2%	47,500		
					Subtotal Indirect Costs	**$729,000**	**23.5%**
					Total Estimated Project Costs	**$3,104,000**	

Operation and Maintenance Cost Estimate

1 Labor Expense					
Operating Labor	hours	1,044	$35.00	36,540	
Maintenance Labor	hours	1,392	$30.00	41,760	
				Subtotal Labor Expense	**$78,000**
2 Energy Costs					
Process Energy	kwhr	290,000	$0.07	20,300	
Raw Water Pumping	kwhr	1,200,000	$0.07	84,000	
Backwash Pumping	kwhr	20,000	$0.07	1,400	
				Subtotal Energy Cost	**$106,000**
3 Chemical Costs					
Ferric Chloride	lbs	33,790	$0.40	13,516	
Chlorine	lbs	16,895	$1.20	20,274	
Polymer	lbs	1,689	$3.00	5,068	
				Subtotal Chemical Costs	**$39,000**
4 Sludge Disposal					
Pumping Costs	kwhr	1,800	$0.07	126	
Wastewater Disposal Costs	lbs	33,800	$0.35	11,830	
				Subtotal Sludge Disposal	**$12,000**
5 Maintenance Materials					
0.5% of Construction Costs per Month	Month	12	$990		**$12,000**
6 Administration and Lab Costs					
	Month	12	$1,750		**$21,000**
Total Annual Estimated Operations and Maintenance Costs					**$268,000**
			Contingency	30%	$80,000
				Total Annual O & M Costs	**$348,000**

Estimated Cost of Water Production

Total Estimated Capital Cost				$3,104,000
Amortized Capital Cost (20 yrs @ 8%)				$316,100
Annual O & M Costs				$348,000
			Total Annual Estimated Costs	$664,100
Maximum Water Production Rate	**1890gpm**	**2.70MGD**		
Efficiency	97%			
Percent of Capicity Used	70%			
Average Daily Production (1,000 gal.)	1,850			
Total Annual Production (1,000 gal.)	675,250			
			Operations and Maintenance Cost per 1,000 Gallons	**$0.52**

Capital Cost per 1,000 gallons	**$0.47**
Total Cost per 1,000 gallons	**$0.98**

Design Criteria

Plant Flow	mgd	2.70
Hydraulic Loading Rate	gpm/sf	**5**
Pressure Vessel Surface Area	sf	375.0
Number of tanks		4
Tank Diameter	ft	5.5
Building Size	sf	1268
Bldg Passage Way	ft	3

Table E.2
Detailed cost analysis for 5.4 mgd plant capacity

Lakewood, CA Created: January 4, 2004
Arsenic and Manganese Removal Facility Revision: 1
Arsenic Removal - Ferric Chloride and Pressure Filter- Plant Capacity -5.4 MGD

Item	Description	Units	Quant.	Unit Cost	Subtotal	Total Cost Yr 2004 Dollars
	Construction Cost Estimate					
1	**General Administration**					
1.01	Mobilization/Bond/Insurance	ls	1	160,000	160,000	
1.02	Environmental Protection	ls	1	30,000	30,000	
1.03	Plant Startup	ls	1	50,000	50,000	
1.04	Record Drawings	ls	1	15,000	15,000	
					Subtotal General Administration	**$255,000**
2	**Yard Piping, Site Grading, and Restoration**					
2.01	Site Preparation	ls	1	10,000	10,000	
2.02	Yard Piping	lf	800	45	36,000	
2.03	Earthwork	ls	1	16,000	16,000	
2.04	Paving	sy	1,950	29	56,550	
2.05	Restoration and Landscaping	ls	1	20,000	20,000	
					Yard Piping, Site Grading, and Restoration	**$138,550**
3	**Treatment Plant Building**					
3.01	Plumbing Allowance	ls	1	18,000	18,000	
3.02	Process Piping	ls	1	50,000	50,000	
3.03	Static Mixer	ls	4	8,000	32,000	
3.04	Backwash Drain Piping	ls	1	30,000	30,000	
3.05	Pressure Filters/Media/Automated Valves	ea	4	230,000	920,000	
3.06	Finishes	ls	1	45,000	45,000	
3.07	Grating	ls	1	9,000	9,000	
3.08	Misc Structures	ls	1	9,000	9,000	
3.09	Misc. Metals	ls	1	9,000	9,000	
3.10	Instrumentation and Controls	ls	1	120,000	120,000	
3.11	Telemetry Programming	ls	1	110,000	110,000	
3.12	Electrical and Transformer	ls	1	237,000	237,000	
3.13	Standby Power	ls	1	110,000	110,000	
3.14	HVAC Allowance	sf	4,000	5	20,000	
3.15	Building	sf	4,000	105	420,000	
					Subtotal Treatment Plant Building	**$2,139,000**
4	**Chem Feed and Backwash System**					
4.01	Ferric Chloride Feed System	ls	1	50,000	50,000	
4.02	Polymer Feed System	ls	1	35,000	35,000	
4.03	Hypochlorite Feed System	ls	1	50,000	50,000	

4.04	Chemical Feed Piping	ls	3	8,000	24,000	
4.05	Lab and Office Equipment	ls	1	9,500	9,500	
4.06	Recycle Pump Sta	ls	1	50,000	50,000	
4.07	Backwash Storage Basin	ea	4	55,000	220,000	
4.08	Backwash Blowers and Piping	ea	4	14,000	56,000	
4.09	Electrical & Controls	ls	1	100,000	100,000	
					Subtotal Backwash Modifications and Equipment	**$594,500**
					Construction Subtotal	**$3,127,100**
			Contingency	30%	938,000	
					Estimated Construction Costs	**$4,065,000**
			Estimated Indirect Costs			
			Geotechnical Engineering		30,000	
			Surveying		40,000	
			Property Acquistion (Acres)	2	100,000	
			Engineering Report		30,000	
			Design Engineering	14%	569,100	
			Construction Services and Startup	7%	284,600	
			Administration	2%	81,300	
			Legal and Permits	2%	81,300	
					Subtotal Indirect Costs	**$1,216,000**
					Total Estimated Project Costs	**$5,281,000**

Operation and Maintenance Cost Estimate

1	**Labor Expense**					
	Operating Labor	hours	1,392	$35.00	48,720	
	Maintenance Labor	hours	1,670	$30.00	50,112	
					Subtotal Labor Expense	**$99,000**
2	**Energy Costs**					
	Process Energy	kwhr	590,000	$0.07	41,300	
	Raw Water Pumping	kwhr	2,400,000	$0.07	168,000	
	Backwash Pumping	kwhr	40,000	$0.07	2,800	
					Subtotal Energy Cost	**$212,000**
3	**Chemical Costs**					
	Ferric Chloride	lbs	67,579	$0.40	27,032	
	Chlorine	lbs	33,790	$1.20	40,547	
	Polymer	lbs	3,379	$3.00	10,137	
					Subtotal Chemical Costs	**$78,000**
4	**Sludge Disposal**					
	Pumping Costs	kwhr	3,500	$0.07	245	
	Wastewater Disposal Costs	lbs	67,600	$0.35	23,660	
					Subtotal Sludge Disposal	**$24,000**
5	**Maintenance Materials**					
	0.5% of Construction Costs per Month	Month	12	$1,690		**$20,000**
6	**Administration and Lab Costs**					

Month	12	$3,000	**$36,000**

Total Annual Estimated Operations and Maintenance Costs		**$469,000**
Contingency	30%	**$141,000**
Total Annual O & M Costs		**$610,000**

Estimated Cost of Water Production

Total Estimated Capital Cost	$5,281,000
Amortized Capital Cost (20 yrs @ 8%)	$537,900
Annual O & M Costs	$610,000
Total Annual Estimated Costs	**$1,147,900**

Maximum Water Production Rate	**3780gpm**	**5.40MGD**
Efficiency	97%	
Percent of Capicity Used	**70%**	
Average Daily Production (1,000 gal.)	3,700	
Total Annual Production (1,000 gal.)	1,350, 500	

Operations and Maintenance Cost per 1,000 Gallons	**$0.45**
Capital Cost per 1,000 gallons	**$0.40**
Total Cost per 1,000 gallons	**$0.85**

Design Criteria

Plant Flow	mgd	5.40
Hydraulic Loading Rate	gpm/sf	**5**
Pressure Vessel Surface Area	sf	750.0
Number of tanks		6
Tank Diameter	ft	5.2
Building Size	sf	1737
Bldg Passage Way	ft	3

REFERENCES

Amy, G., M. Edwards, M. Benjamin, K. Carlson, J. Chwirka, P. Brandhuber, L. McNeill, and F. Vagliasindi. 2000. *Arsenic Treatability Options and Evaluation of Residuals Management Issues*. Awwa Research Foundation, Denver, CO.

Chang, S. D., W. D. Bellamy, H. Ruiz, and C.W. Spangenberg. 1994. *Arsenic Removal by Enhanced Coagulation and Membrane Technology*. 1994 AWWA Annual Conference, New York, NY.

Chiu, V. Q. a. J. H. 2000. "Arsenic Adsorption and Oxidation at Maganite Surface. 1. Method for Simultaneous Determination of Adsorbed and Dissolved Species." *ES&T* 34(10): 2029-2034.

Clifford, D., G. Ghurye, and A.Tripp. 1998. Arsenic Ion-Exchange Process with Reuse of Spent Brine. *1998 AWWA Annual Conference*. Dallas, TX.

Dreihaus, W. R. S. M. J. 1995. "Oxidation of Arsenate (III) with Manganese Oxides in Water Treatment." Water Research 29(1): 297-305.

Edwards, M. 1994. "Chemistry of Arsenic: Removal During Coagulation and Fe-Mn Oxidation." *JAWWA* 86(9): 64-78.

Edwards, M.A., S. Patel, L. McNeill, H. Chen, M. Frey, A.D. Eaton, R.C. Antweiler, and H.E. Taylor. 1998. Considerations in As Analysis and Speciation. *Jour. AWWA*, 90:3:103.

Ferguson, J. and J. Gavis 1972. "A Review of the Arsenic Cycle in Natural Waters." *Water Research* 6: 1259-1274.

Frey, M.M., D.M. Owen, Z.K. Chowdhury, R.S. Raucher, and M.A. Edwards. 1998. Cost to Utilities of a Lower MCL for Arsenic. *Journal AWWA*, March, pp. 89-102.

Fruth, D.A., J.A. Drago, M.W. Tikkanen, K.D. Reich, L. Lund, S.B. Nielsen, L.Y.C. Leong. 1996. A River Basin Based Method to Estimate Arsenic Occurrence in California Surface Waters. AWWA WQTC, Boston, MA.

Hering, J. and M. Elimelech 1996. *Arsenic Removal by Enhanced Coagulation and Membrane Processes*. Denver, CO, AwwaRF.

Hering, J.G., P.Y. Chen, J.A. Wilkie, M. Elimelech, and S. Liang. 1996. Arsenic Removal by Ferric Chloride. *Jour AWWA*, 88:4:155.

Ghurye, G. and D. Clifford. 2000. Laboratory Study on the Oxidation of Arsenic III to As V. *AWWA Water Quality Technology Conference Proceedings*, American Water Works Association, Denver, CO.

Knocke, W. R.; Van Benschoten, J. E.; Kearney, M. J.; Soborski, A. W.; and Reckhow, D. A. 1991. Kinetics of Manganese and Iron Oxidation by Potassium Permanganate and Chlorine Dioxide, *Journal of the American Water Works Association*, 83[6], 80-87.

Lauf, G. F. and K. S. Plsarczyk. 1995. Optimization of Manganese Greensand for Arsenic Removal. 1995 AWWA WQTC, New Orleans, LA.

Lauf, G. F. and M. A. Waer. 1993. *Arsenic Removal Using Potassium Permanganate*. 1993 AWWA WQTC, Miami, FL.

Magyar, J. 1992a. "Arsenic Removal Study." *Saskatchewan Operators Clearwater Reporter* 16(4): 1-2.

Magyar, J. 1992b. *Kelliher Arsenic Removal Study*, Saskatchewan Environment and Public Safety.

Masscheleyn, R.D. Delaune, and W.H. Patrick, Jr. 1991. Effect of Redox Potential and pH on Arsenic Speciation and Solubility in a Contaminated Soil. *Environ. Sci. Technol.*, Vol. 25, No. 8, pp. 1414-1419.

McNeill, L. S. and M. Edwards. 1994. *Arsenic Removal in Full-Scale Treatment Plants: Theory versus Practice*. 1994 AWWA WQTC, San Francisco, CA.

McNeill, L. S. and M. Edwards. 1995. "Soluble Arsenic Removal at Water Treatment Plants." *JAWWA* 87(4): 105-113.

McNeill, L. S. and M. Edwards. 1996. *Predicting Arsenate Removal During Metal Hydroxide Precipitation*. 1996 AWWA Annual Conference, Toronto, Ontario, AWWA.

McNeill, L. S. and M. Edwards. 1997. "Predicting as Removal: During Metal Hydroxide Precipitation." *JAWWA* 89(1): 75-86.

Welch, A.H., Westjohn, D.B., Helsel, D.R. and Wanty, R.B., 2000. Arsenic in ground water of the United States: Occurrence and geochemistry. Ground Water 38(4) 589-604.

ABBREVIATIONS

ACWA	Association of California Water Agencies
As(III)	arsenite
As(V)	arsenate
AwwaRF	Awwa Research Foundation
BWW	backwash waste
EPA	U.S. Environmental Protection Agency
ft^2	square foot
gpm	gallons per minute
m^2/g	square meters per gram
MCL	maximum contaminant level
mg	milligram
MG	million gallons
mg/L	milligrams per liter
Mgd	million gallons per day
NPDES	National Pollution Discharge Elimination System
NTNC	non-transient, non-community
O&M	operation and maintenance
PFT	paint filter test
PLC	programmable logic controller
RCRA	Resource Conservation and Recovery Act
rpm	rotations per minute
SCWC	Southern California Water Company
STLC	Soluble Threshold Limit Concentration
TBLL	Technically Based Local Limits
TCLP	Toxicity Characteristic Leaching Procedure
TOC	total organic carbon
TTLC	Total Threshold Limit Concentration
UFRV	uniform filter run volume

WET	Waste Extraction Test
μg/L	microgram per liter

IWA Publishing's authorised EU representative for General Product Safety Regulations is Diane D'Arras, 15 rue Duret, 75116 Paris, France, e-mail: safety@iwap.co.uk.

Printed and bound by CPI Group (UK) Ltd, Croydon, CR0 4YY
17/09/2026
02218808-0014